THE CLIMATE CHANGE GARDEN

THE CLIMATE CHANGE GARDEN

Down to Earth Advice for Growing a Resilient Garden

SALLY MORGAN & KIM STODDART

Brimming with creative inspiration, how-to projects, and useful information to enrich your everyday life, Quarto.com is a favorite destination for those pursuing their interests and passions.

First Published in USA in 2023 by Cool Springs Press, an imprint of The Quarto Group, 100 Cummings Center, Suite 265-D, Beverly, MA 01915, USA.
T (978) 282-9590 F (978) 283-2742 Quarto.com

27 26 25 24 23 1 2 3 4 5

ISBN: 978-0-7603-7948-6

Digital edition published in 2023
eISBN: 978-0-7603-7949-3

Library of Congress Cataloging-in-Publication Data available

Original Design: Jayne Jones
Design (Current Edition): The Quarto Group
Page Layout: Megan Jones Design
Photography: Kim Stoddart, Sally Morgan, and Shutterstock
Front Cover Illustration: Sue Gent

Printed in China

CONTENTS

Foreword: Why It's No Longer Gardening as Usual

Greater extremes of weather have been impacting communities across the world for some time, yet the increasingly violent storms of the past few years have "flipped the switch" on the challenging reality that we now face.

As gardeners, we are often more directly aware of the seasons than many, being tuned to the natural rhythms of germination, growth, fruiting, harvesting, death, and renewal of the plants and produce that we tend. We're out in the elements and see firsthand how the weather directly impacts our gardens. While every year is different, there has always been a comfort to the regular and predictable changing of the seasons and the routines and practices that we have carried out because that is just what we've done at that time every year.

Yet there's no getting away from it: The climate and weather patterns of the past are changing fast, and our gardening practices need to adapt to catch up. Worldwide, the lack of global leadership on environmental matters continues to frustrate and cause a sense of powerlessness among individuals. As

Floods may become more frequent.

A changing climate may bring more droughts.

gardeners, we already recognize that there's no longer going to be traditional seasons and growing methods, and most of us are extremely worried and looking for answers. Our book aims to provide just that.

After talking about this subject a lot over the years—in Kim's case in the gardening pages of UK publications such as *The Guardian* and teaching more resilient techniques in her courses, and in Sally's case writing on food, farming, and environmental matters—we realized there was a pressing need for an accessible "how to garden in a changing climate" guide that we, with our respective experience on the subject, could provide.

Of course, the uncertainty of the future we now face is intimidating. Yet, learning how to protect our precious outside space against extreme rain, sunshine, wind, snow, and goodness knows what else that is ahead of us will be key. As will knowing which plants are best suited to deal with such extremities in the first place; the techniques, practices, and equipment that can be employed in an existing space; and the future designs that can help provide a greater robustness. All of this and much more is covered in this book, alongside lots of practical take-home advice and inspirational ideas to help you on your way to more resilient gardening.

Kim and Sally

aka the Climate Change Gardeners
(climatechangegarden.uk)

Glossary of Terms for Global Readers

We know gardening terminology varies from one region to the next and that can sometimes be confusing. We are United Kingdom–based authors, but many of our readers are located in the United States and even in other parts of the world. Throughout this book, U.S. gardening terminology is used. However, we thought a brief glossary of some of the most frequently used terms would be helpful.

U.S. Term	U.K. Term
Arugula	Rocket
Beet	Beetroot
Boxwood	Box
Downspouts	Downpipes
Drain field	Soakaway
Eggplant	Aubergine
Gasoline	Petrol
Granite block	Granite-sett
Pieces of timber or lumber	Sleeper
Planting hole for seeds	Drill
Potting soil or potting mix	Potting compost
Rain barrel	Rain butt; water butt
Row cover	Fleece
Rutabaga	Swede
Shredded leaves	Shreddings
Small seed inside a fruit	Pip
Waterway	Water course
Water well	Bore hole
Zucchini; summer squash	Courgette

INTRODUCTION

A Taste of Things to Come?

Extreme weather has been seen worldwide, and 2021 and 2022 were arguably the years the realities of climate change warming finally broke through into public and political consciousness. From winter storms in Texas in February, when temperatures dropped to 8°F (-13°C) in some areas and 3.5 million homes and businesses were left without power, to Hurricane Ida, which hit the United States in late summer and caused estimated economic losses of $65 billion, to flooding events in the United Kingdom; China; New South Wales, Australia; and record-breaking heatwaves across Europe and the UK in 2022, the financial impact has been dramatic.

Drought caused Lake Mead behind the Hoover Dam, the largest reservoir in the United States, to fall to record low levels in 2022.

In recent years we've also seen record temperatures in the northwest United States and China; and wildfires across Australia, Oregon, California, and Greece. Scientists report that the western United States is experiencing the worst megadrought (a prolonged period of dryness spanning 20 years) in 1,200 years and predict even drier decades to come. Also in the United States, 2021 was overall the fourth warmest year since records began 127 years ago. Not that shocking, until you consider that the six warmest years on record have all occurred since 2012. The list goes on.

From a gardening perspective, changing extreme weather events put further pressure on more delicate plantings, and gardeners themselves feel the stress as they try desperately to keep up with whatever is coming next. Building resilience to our changing climate and shoring up our whatever-the-weather defenses is key.

What Do We Know?

The global rise in the levels of carbon dioxide is well documented; it's been happening since the start of the industrial revolution, when we began to burn coal. Our addiction to fossil fuels, coupled with deforestation and changes in land use and farming, has boosted the levels of carbon dioxide in the atmosphere from 280 ppm in 1750 to 414 ppm in 2021, despite a temporary reduction in global emissions during the COVID pandemic in 2020. The current rate of increase is four times that of the 1960s.

Carbon dioxide, together with water vapor, methane, ozone, CFCs, and nitrous oxide, are called greenhouse gases (GHG) because they trap heat in the atmosphere and cause temperatures to rise, just like a closed greenhouse on a hot day. It is GHGs that are behind global warming.

According to the Intergovernmental Panel on Climate Change (IPCC): "Human activities are estimated to have caused approximately 1.8°F (1.0°C) of global warming above pre-industrial levels, with a likely range of 1.4°F to 2.2°F (0.8°C to 1.2°C). Global warming is likely to reach 2.7°F (1.5°C) between 2030 and 2052 if it continues to increase at the current rate." Also, different parts of the globe are experiencing more warming than others. For example, it happens more over land than water and it's up to three times greater in the Arctic.

Between 1998 and 2010, the rises in average surface temperatures were not as high as predicted because weather systems and events, such as volcanic eruptions throwing clouds of small particles into the atmosphere and the cooling effect of La Niñas (see page 18), helped offset human-made changes. But now scientists warn of the opposite and predict a peak warming phase.

Forest fires raged across California in 2021.

Action is needed, and fast, but the trouble is, it's now difficult to stop these changes. We can only hope to slow them down.

What Is Likely to Happen Weatherwise?

More heat waves in Europe, more fires in the United States, and more droughts in Australia and South Africa are predicted. As the oceans warm up more quickly than the air above them, more energy is picked up by the oceanic weather systems, causing more hurricanes and typhoons and, as a result, more flooding. James Renwick of the Victoria University of Wellington in New Zealand said, "If the warming trend caused by greenhouse gas emissions continues, years like 2018 will be the norm in the 2040s and would be classed as cold by the end of the century."

We have to be prepared for more erratic weather—more storms, high winds, droughts, and extreme high or low temperatures.

Will we see more residential flooding?

LEARNING FROM THE PAST:
Past Changes

Climate change is not new. Let's take the United Kingdom as an example. It has had periods of warmer weather before. The Romans used to grow grapes and lentils in Britain, and there was another warm period about a thousand years ago in medieval Britain. Meanwhile, the Little Ice Age (sixteenth to nineteenth centuries) was a time of cooling and extreme winter weather, with snow on the ground for many months, frequent storms, and cold summers. It was cold enough for rivers to freeze over, and to be used for frost fairs. The growing season was shorter, crops failed, and famine was common. Gardeners adopted the idea of the walled garden to create a microclimate that suited growing vegetables.

Adaptation is going to be key.

A gravel rockery in Kim's climate change garden.

Extreme Weather Gardening

In an established garden, the problem is not that the weather is too cold or too warm, or that spring is early, or that frost comes late. It's the fact that all these increasing extreme-weather variables are tending to occur within a short space of time and garden plants can't cope. It takes a resilient plant to survive all that nature throws at it, and in a climate-changed world, it's not just the long-term rise in temperatures but also the extreme weather events that are so disruptive for plants. These stresses make them more vulnerable to pests and disease.

Kim's naturalistic gardens work with nature, rather than trying to meticulously control it.

It's no longer gardening as usual. We're already seeing changes—the earliest spring, the mildest winter, the wettest year. New pests and diseases are being reported every year. Gardens take years to mature, so if our gardens have any chance of surviving the future climate in ten, twenty, thirty, or even eighty years, we need to start implementing plans for change right now.

Urban gardens are likely to be leading the change, as towns and cities will see greater shifts in weather patterns due to the "heat island" effect caused by the presence of streets, buildings, and traffic, which all generate heat. There will be new challenges too, such as coping with heavy downpours that create flash floods and waterlogging.

It's going to be hard, and as gardeners, we are going to have to make uncomfortable changes and maybe lose some of our best-loved features. Which of our favorite garden trees will be able to survive the buffeting of storms or the stress of drought? Will you even want a lawn? Will the classic cottage garden survive? What about our rose gardens? Head gardeners in charge of major gardens are already adapting and planning for the future, so you should too.

Yet it's not all doom and gloom, so although we outline the challenges in the pages ahead, please be aware that the changes offer opportunities too. No more mowing the lawn if gravel gardens take pride of place instead, and exotic tropical plants may fill the flower borders. You may lose a few favorites but gain new ones—how about growing eggplants, chickpeas, soybeans, and lentils in the vegetable plots and palms in the flower beds?

You will need to listen to your garden—pay attention to what is happening now, learn from hard lessons, and try to work out how to best cope with the many changes that are to come. How are you going to adapt your garden to climate change? Grandparents are planting trees for their grandchildren to enjoy. That's a lovely sentiment, but what species should they be planting? What should you be planning in your garden for ten, twenty, and even thirty years ahead?

LEFT: **Will our lawns weather out the climate change storm?**

What Causes Unusual Weather Patterns?

Weird weather patterns are often linked to solar activity. European weather is very dependent on the jet stream. This is a jet of warm air that blows from west to east, across the Atlantic, giving northwest Europe its mild climate. But a period of low solar activity throws the jet stream off its normal pattern. It may shift north or south of its usual route and cause what meteorologists call a sudden stratospheric warming (SSW). In the weeks after an SSW, northern Canada, Alaska, the Middle East, and Central Asia may experience unusually warm weather, while Siberia, northern Europe, and the continental United States get extreme cold weather, as happened in Europe in 2018, when the "Beast from the East" caused vast snowstorms across the region.

Other important weather factors are El Niño and La Niña years. These are regular shifts in patterns of weather in the eastern tropical Pacific Ocean that occur every few years and have effects that are felt throughout the world. The 2015 El Niño effect was one of the strongest ever seen, with massive ocean warming. The effect usually lasts for about a year; it warms up the surface of the Pacific and weakens the trade winds, causing changes in wind patterns and rainfall. The extra energy from the warmer ocean heats up the atmosphere, which creates more intense weather events. This leads to increased rainfall in South and North America and below-average temperatures. La Niña is the opposite, as the ocean waters of the eastern Pacific are cooler and the trade winds are stronger. This results in cooler, drier weather in the tropical eastern Pacific and can even lead to an increased risk of colder winters in northern Europe.

Country	Predicted Changes
Australia	Higher temperatures, more drought, fire seasons, and flooding are being seen. Soil is already dry but getting even drier still. Damaging hailstorms are becoming more prevalent, and gardeners are seeing more intense, less predictable weather events, such as high winds and heavy downpours. The average temperature is predicted to increase 3.6°F (2°C) above 1990 levels by 2030, and as much as 10.8°F (6°C) by 2070. The Commonwealth Scientific and Industrial Research Organization (CSIRO) forecasts that the next few decades will see further temperature increases, more extremely hot and fewer extremely cool days, sea level rises, more frequent and longer-lasting heat waves, and a decrease in cool-season rainfall across southern Australia. They predict more intense rainfall and an increase in the number of high fire weather danger days over a longer period of time for southern and eastern Australia in particular.
Canada	The Atlantic Coast of Canada is the most threatened, especially the coastal communities where there is the potential for more storm intensity and rising sea levels from Arctic ice melt and storm surfaces. The central farming belt may see lower summer rainfall, retreating glaciers, and more drought overall, while the Pacific Coast is at risk of more storms, forests fires, landslides, and flooding.
United Kingdom	Already warmer by about 1.5°F (0.8°C) compared with 1961 to 1990, the United Kingdom is forecast to see more frequent heat waves in summer, maybe as much as twice a decade, compared with twice every hundred years in the early 2000s. By 2040, severe heat waves may occur every other year. The higher-than-average temperatures will mean a longer growing season. An increase in average temperature of just 1.8°F (1°C) will extend the growing season by three weeks in the south of England and by ten days in the northwest. Winters are likely to be milder and wetter and feature more storms. If temperatures were to rise by 3.6°F (2°C), southern England would experience a climate more like southern France, while a massive 7.2°F (4°C) rise would result in a climate like that of southwest Portugal. By the year 2050, much of the United Kingdom will suffer from significant water shortages.

Country	Predicted Changes
United States	Northern areas are already wetter, especially in winter and spring, while southern areas are much drier overall, especially the southwest. They are seeing more hurricanes with greater intensity (more category 4 and 5). The difference in temperatures between the north and south is becoming smaller as the jet stream weakens. By 2035, the northeast is set to be 3.6°F (2°C) warmer on average as the region warms faster than any other region in the Lower 48, double the rates of elsewhere. This means less snow, warmer winters, and warmer coastal waters, all of which impacts the maritime climate overall. The region is influenced significantly by a warmer Arctic, with loss of ice cover causing more ice melt. Sea level rises off the coast are already rising faster than in the rest of the United States. Boston, for example, has seen an 8-inch (20 cm) increase in sea levels since 1950, while the Gulf of Maine is warming faster than 99 percent of the world's oceans overall, which is disrupting fisheries and affecting the maritime climate dramatically. Weather systems are becoming more persistent as they get "stuck," and therefore hang around for longer than they have previously. Summer days are getting warmer, with temperatures of 77°F (25°C) and above more frequent, and tropical nights with balmy weather of 68°F (20°C) and above. By 2100, the Connecticut climate in summer could be the same as present-day South Carolina, with temperatures exceeding 100°F (38°C) for twenty-eight days a year or more. The Midwest is already seeing warmer winters with persistent and substantially higher temperatures combined with longer periods of high humidity. The region can expect more snowfall and more frequent and intense summer thunderstorms too. Yet with variable weather patterns, there may still be the occasional severe winter, as it's not just a case of warming overall.

TOO MUCH WATER

Does your garden cope well with a heavy and prolonged deluge of rain? When people think of climate change, they anticipate hotter summers and potential water shortages but, perhaps surprisingly, a distinct excess of water in the garden will be one of the main problems in our climate-changed future, alongside coping with summer drought. And more rain will fall during intense weather events. If you've experienced flooded beds and a sodden lawn recently, then it's really time to get prepared.

Throughout this chapter we'll keep coming back to the basic principles used by landscape architects and permaculturalists—slow it, spread it, and sink it. In other words, slow down the movement of water, allow it to spread out, and create opportunities for the water to drain into the soil, where it will be stored for future use. Any excess water that cannot be held by the soil can then be collected in ponds and tanks.

LEFT: Flooded vegetables in a walled garden.

We're Expecting More Rain

Rainfall patterns are changing. Many places have experienced record-breaking wet years. For example, the United Kingdom has experienced seven of the ten wettest years on record since 1998, while in March 2021 the coastal zone of New South Wales in Australia had the wettest week since 1900. But we are likely to experience more extreme rainfall events with more heavy, monsoon-like downpours. On July 20, 2021, Zhengzhou, China, received a record-breaking 8 inches (20.2 cm) of rain in just one hour, which overwhelmed the city's storm drain system and flooded roads and subways; in Germany in July 2021, 6⅓ inches (16.2 cm) of rain fell in one day, resulting in huge floods that took the lives of 179 people.

Tens of millions of homes around the world are at risk of flooding from rivers, coastal, or surface waters. With more rain in the cards, combined with the predicted rise in sea levels, flooding is going to become even more commonplace.

Floodwaters around raised vegetable beds in winter.

Did You Know?

A lot of water falls on a roof in a year. You can calculate just how much by multiplying the annual rainfall (mm) by the roof surface area (m^2) to find the volume in liters that will run off the roof. Let's assume your average rainfall is 965 mm. For 100 m^2 of roof, that's a massive 96,500 liters of water a year. (For imperial measurements, change the square footage of the roof into square inches, then multiply by the annual rainfall in inches, then divide by 231 [because 1 gallon = 231 cubic inches]. So for this example, 154,921 square inches of roof × 38 inches of rainfall ÷ 231 = 25,484.8 gallons.

TIP

Heavy summer storms can do untold damage to herbaceous plants that are laden with leaves and flowers, so the plants will need plenty of staking to avoid damage.

Going with the Flow

Part of the problem comes from the fact that water levels can rise very quickly. After heavy rain, a trickle in a stream soon becomes a torrent and can flood nearby gardens. Even quite small changes, such as a blocked drain or culvert, can lead to flooding. Extensions to buildings, new walls, and garages can cause problems as well. For example, a new wall can block the route that water used to take, so after a heavy deluge, water has nowhere to go and floods the ground in front of the wall. Your garden may simply be lower than neighboring gardens or the road, or be located at the bottom of a hill, so water flows in. Unless your neighbors install drainage and divert the water running off their garden into yours, you are going to have problems. But in the future, your garden may flood after heavy rain simply because it, or the surrounding area, can't cope with the volume of water.

Flood Warnings—When Your Garden Is at Severe Risk

It has been raining heavily for a week and you are sent an alert that your home lies in an area that is expected to severely flood. What can you do? Local authorities may distribute sandbags around at-risk areas, but if you live in a water-vulnerable location, it may be wise to have your own stockpile of sandbags and flood boards on hand. You can use them to protect areas such as the base of a greenhouse and its glass panels, or to direct water away from a weak wall. Flood boards can also be positioned across a gate to prevent floodwater from entering a garden.

TIP

If you have lots of wooden structures in your garden, remember that heat and moisture are the two factors that will increase wood rot. If wood needs replacing, think about replacing it with metal or more sustainable materials, such as recycled plastic.

If you have time, you can also prepare by doing the following:

- Lift pots, bags of soil and compost, and equipment off the floor of the greenhouse, garage, and shed.
- Move valuable equipment, such as a lawn mower, onto pallets.
- Unplug electrical equipment, such as heaters, lights, and pond pumps.
- Turn off the water supply to your garden.
- Make sure containers of fuel, oil, and pesticides are placed on shelves out of reach of any floodwaters and that gas cylinder valves are turned off.
- Move pots and garden ornaments onto walls or higher ground.
- Secure or weigh down fruit cages, cold frames, chairs, and play equipment.
- Place netting over any pond and secure it tightly so your fish are not washed away.
- Weigh down manhole covers with sandbags to prevent them from lifting during flooding and creating a trip risk.
- Harvest any crops in the ground.

If you are unlucky and your garden is severely flooded, don't forget to take photos for your insurance claim and don't throw anything away until the insurance assessor has visited.

Cleaning Up

A garden can look a complete mess after surging floodwaters have passed through. It wouldn't be so bad if it were just river water, but most floodwater carries with it all sorts of contaminants, including plastics, sewage, manure, and chemicals, such as oil, pesticides, and more. So, if you suspect that the water is contaminated, wear protective clothing and waterproof boots while in the garden and keep pets away.

It's best to avoid going into the garden until the floodwaters have drained away. Before you do so, make sure the electricity supplying any outside sockets is turned off and don't turn it back on until everything has been checked. All the rubbish from a flooded garden may be contaminated, so it must be disposed of in the correct way as guided by the local authorities.

Quick Replacement Vegetable Beds

If your vegetable plot has been flooded and you have lost your crops, the quickest way to recover is to build some new raised beds and fill them with bought-in soil and compost so that you can still make use of the growing space. The ultimate quick fix raised bed is a bulk bag filled with compost. You can also make use of containers and grow bags. But don't be tempted to fill the new beds with compost from any compost bins that have been flooded.

Although it may be tempting to keep sandbags for another time, they should also be thrown away, as they will be contaminated. The same is true for any bags of compost, sand from play areas, bark mulch, and so on that came into contact with floodwater. Never, ever eat any vegetables that were in the ground at the time of flooding, even those that would normally be cooked. Guidance suggests that the ground should not be used for at least one year—and even longer for salad crops—to ensure that sure any contaminants, spores, or disease-causing bacteria are long gone.

Water can also do a surprising amount of damage to the foundations of buildings and retaining walls, so check for damage. Unblock drains and hose down hard surfaces, such as patios, paths, and walls. Then check for any cracks or other signs of damage. Don't forget that wooden arches and pergolas may be damaged at ground level.

The soil is going to be waterlogged for some time and, when it is in this state, it is easily compressed and susceptible to compaction, so don't walk on it and cause further harm. If you need to gain access or cross a flower bed, use a wooden board to spread your weight. And don't forget that lawns, too, will be waterlogged, so don't be tempted to walk on them.

After the Flood

If it's a flash flood, it's less likely that there will be long-term damage, but heavy rain or floodwater pouring in from rivers may result in standing water in your garden for days or even weeks. The excess water drains slowly into the soil and creates waterlogged conditions that can stress plants, but in a different way than heat. The water fills the air spaces in the soil, pushing out the air and creating anaerobic conditions around plant roots, which can cause them to die. Few species can cope with waterlogged roots for long, apart from those adapted to living in boggy areas, such as willow.

In the weeks following a flood, look for signs that flooding has damaged your plants; you may notice stunted growth, yellowing leaves, and leaf drop. Leaves wilt because the roots are dying and starting to rot. If you dig up the plant, you will probably see blackened roots and they may smell or have rotted away already. The dying roots mean there is no transpiration occurring, so water and nutrients are not moving around the plant, causing the leaves to wilt. It may seem odd that the plant is water-hungry when it's surrounded by water, but in such circumstances the plant has no means of taking it up. You may also notice that bark starts to peel on shrubs and trees, growth in spring is slow or stunted, and some branches start to die back. Prolonged waterlogging may result in the decay of the root systems of herbaceous plants, so they don't reappear in spring, and bulbs may simply rot in the ground. A standing flood in summer will have more serious consequences than a winter one, as the plants will be at their most active and are therefore less able to cope with the floodwater.

Helping Your Plants

First, cut away any dead stems and branches, and prune damaged plants. Once you see some new growth, give the plant an organic fertilizer. Any valuable plants can be gently removed from the ground and their roots washed. Then replant them in a drier part of the garden or a large pot. If you have a lot of plants to rescue but you still have a waterlogged garden, find the driest area, dig a trench, and backfill it with a free-draining mix of soil, grit, and compost and use it as a nursery bed while you sort out the rest of the garden. For plants that are too large to move, dig a shallow trench around them to help the water drain away from their crown. Forking the ground around shrubs and trees helps boost drainage too.

Floodwaters may have carried away some of the soil's nutrients, especially soluble nitrogen, so to repair the damage, mulch your beds with compost to boost nutrient and organic matter levels and, if necessary, feed the plants in spring. A long-term slow-release organic feed can help trees and shrubs. Don't forget that plants that have damaged roots are also going to be more susceptible to drought, so they will need plenty of water during dry spells. Damp conditions can persist for some time, so expect more slugs and snails, and fungal disease. Eventually, once the soil is dry, you can start remedial action to get rid of any compaction, such as by forking or otherwise loosening the soil.

TIP

A really simple but effective way to trap and slow down water is for everybody in the street to simply put a load of buckets on their yard or patio to catch water. They can then empty them gradually after the rain has stopped or use them to refill rain barrels or cisterns.

If there is little you can do to avoid flooding, aim to grow plants that can cope with floods and waterlogged soil. If you have vegetable beds, focus on planting out in late spring and harvest by late autumn to avoid the wettest months of the year. During the rest of the year, keep the soil covered to protect it and prevent weeds from growing with a layer of compost, black plastic, or mulch. And in the long term, look at some of the slow water options described later, such as building raised beds so that the water runs between the beds, keeping the root zone above the water. Lawns don't thrive with regular flooding and waterlogging, so it may be best to replace a lawn with gravel or a water-tolerant ground cover. You could even plant a bog garden with a raised boardwalk to provide interest and a different vantage point.

Work Together

It's surprising just how much difference a group of neighbors can make when it comes to reducing the risk of flood, so when you can, work together. Joint actions might include collecting rainwater by using water storage tanks, avoiding bare dug-over patches of soil (which don't hold water as well), directing water away from a neighbor's garden, and using some of the options described later to slow the movement of water and reduce the volume of water leaving one garden and entering that of a neighbor.

Preparing for Heavy Rain

In a natural ecosystem, most of the rainwater soaks into the ground. While some is taken up by plant roots, most continues downward to the water table. The water is filtered through the soil and rocks, so clean water recharges the water table. However, in gardens, hard landscaping and other impermeable surfaces mean that water cannot soak into the ground because it has to flow elsewhere.

So, to stop your garden from flooding, you need to slow the movement of water, giving it the chance to soak into the ground or flow gently into local waterways.

The best planning involves observation. On a day when the rain is pouring down, put on your waterproof boots and brave the outside to watch and see firsthand how water moves through your garden. This will give you the best insight possible. Check the following:

- Where does the water run? Does it follow paths and drives, or run across beds and lawns?
- Does water pour in from the street or a neighboring garden?
- Does the water run away or collect into large puddles?
- Are there any gushing downspouts?
- Are there structures in the garden, such as summerhouses and sheds, that are at risk from flooding and need to be moved?

Observe your garden during torrential rain.

- Do you have a sloping outside space? Does your garden slope toward the house or away from it? What are the lowest points in your garden? Does water collect there?
- Are there any local waterways? Is your garden at the lowest point in the area? If so, it's more likely that floodwater from these streams and ditches will flood your garden rather than those of your neighbors.

Armed with this knowledge, you will be able to plan more effectively. Don't forget that you may need to involve your neighbors to help shore up your local flood defenses.

Slowing Down Water

Slowing water means encouraging it to move into the soil and reduce runoff. Thankfully, there are lots of ways you can improve the flood resilience of your garden by incorporating clever water-slowing features. Here are some to think about.

Mulching

Before you start redesigning your garden, think again about your soil. It's such an important natural ally. Spreading a thick layer of compost over the soil each year will boost organic matter and create a permeable surface that water can penetrate and drain through (see chapter 4). Bare soil, devoid of mulch or plants, won't be able to absorb much rainfall and will be more susceptible to nutrient leaching during the winter, so, whatever you do, don't leave it uncovered. Instead, mulch it with compost, cover it with plastic, or grow a green manure. Trees are useful, too, as their roots absorb water from a large area, so a garden with trees and shrubs can absorb more water than the same area of outside space without them.

TIP

Trees are susceptible to flooding and standing water, so new trees can be planted on gentle mounds to help water run away from trunk and roots.

Raised Beds

Raised beds can be incredibly useful in gardens where water collects. In fact, some permaculture gardens flood the land deliberately to bring in nutrients, letting the water run between the raised beds (rather like a traditional water meadow that is flooded in winter). The advantage of a raised bed is that it lies above the water level, so the soil does not get waterlogged. You can read more about this in chapter 5.

The gravel paths and raised beds help make this garden flood resilient.

Impermeable Surfaces

Avoiding large areas of concrete and other impermeable surfaces is important, as they create a lot of runoff. Instead, use paving stones, bricks, or gravel so that water can seep through the gaps. If you do need an area of a solid surface, think about a porous asphalt or permeable concrete so at least some of the water is able to soak in.

French Drains and Weeping Tiles

A French drain is a small trench that is backfilled with gravel. It allows water to drain away from a building, driveway, or lawn, stopping water from collecting and reducing the risk of flash floods. To improve the flow of water further, a perforated drainage pipe known as a weeping tile is laid along the bottom of the trench to help carry water away from the house to an area that can cope with the water. Weeping tiles were once made from terra-cotta, but nowadays they are plastic.

Drains help carry water away from the lawn.

A pebble-lined drainage channel.

Drain Fields

It is usual practice for house downspouts to direct water along French drains into a drain field to allow it to sink into the ground. Builders use a formula to calculate how much water will be pouring off the roof and design the size and length accordingly. A drain field is usually at least 16 feet (5 m) from the house so that water does not drain back toward the foundation. It's typically about 3 feet (1 m) wide and 5 feet (1.5 m deep), backfilled with gravel and positioned so that it is found in the lowest part of the garden where water can flow in and drain away. However, drain fields don't work on heavy clay soil, as the water cannot flow easily through the clay. Any garden structure with a roof, such as a shed or garage, may simply drain onto the ground, so you may need to improve drainage nearby by installing French drains and a small drain field. Water from downspouts should not enter the sewers, as overloading them creates an overflow. If you have downspouts, don't let them discharge near the building, but extend them so they direct the water further away toward grass or flower beds, where the water has a chance to permeate into the ground.

Rain Barrels and Storage Tanks

These structures are great for harvesting water for use in the garden in summer but, once they are full, they are no help at all in slowing down the impact of heavy rainfall. After a storm, when they are filled to capacity, it is best to let the water drain away gradually so that when it rains again, they can collect a large volume of water.

Another option to consider is the use of leaky barrels filled with gravel. The runoff from the roof collects in the barrel, but only drains out slowly. This is called attenuation. While rain barrels and tanks help intercept water pouring off roofs, you need to connect them to a system of drains, so the water is moved to places where it can slowly flow away.

A swale carries away excess water.

A downspout empties into a drain and rain garden.

Swales and Berms

Slopes are vulnerable areas in the garden and it's vital to avoid areas of bare ground. Water moves down a slope at speed, taking the shortest route. It can carry soil with it, causing soil erosion. You may think that soil erosion only occurs on farmland or cleared forest land, but it can happen in your garden too. It's a good idea to try to reduce the slope by terracing and using swales to carry the excess water away. Swales are ditches that are built along the garden contours and are physical barriers to slow the movement of water and runoff. The raised bank beside the swale is called a berm and they work together to slow runoff, with the swale filling with water and slowing its movement. Swales can be built so that they direct water to a rain garden or a storm drain (see chapter 5).

Permeable Paths

Paths can inadvertently become routes for water to rush through the garden and collect in one area. To avoid this, make sure the path has a permeable surface rather than a hard one or follows a more sinuous route, following contours, so the flow of water is slowed by forcing it to take a longer path. With more rain at certain times of the year, paths are more likely to be damp and covered with algae and moss.

RIGHT: **A permeable gravel path.**

Clever design directs water away from the path.

A green roof holds water and releases it slowly.

Sump Pit

If you still get water collecting in your garden, you can build a sump pit at the lowest point. A sump is ideally a 1 square yard (1 m^2) area dug as deep as necessary to let the water drain away. It is often simply backfilled with hard debris, such as old bricks and stones, but a modern technique is to use plastic attenuation crates surrounded by a permeable membrane. The sump is topped with a final layer of gravel. It collects water that has been directed through drainage trenches away from the house.

Green Roofs and Vertical Gardens

Covering surfaces with vegetation will help absorb some of the water and reduce the flow through the downspouts. Nowadays, green roofs are quite common and are generally planted with turf or sedum and other succulents that can tolerate summer drought conditions (see page 93).

Rain Gardens

A rain garden is simply a shallow depression, backfilled with a well-drained soil and filled with plants that can cope with being flooded on a temporary basis. Rain gardens are great at temporarily holding stormwater, slowing down water runoff from hard surfaces and, as a bonus, they act as a natural filter. The best place to build your rain garden is several yards (meters) away from the house on a flat area or one with only a gentle slope (less than 10 degrees). Swales and downspouts direct the water away from the house to the rain garden, where it collects and from there drains into the ground or is directed by an overspill area or weeping tile to another area of the garden. For obvious reasons, a rain garden needs to be located far away from septic tanks and wells. For more details, see page 91.

CASE STUDY

KIM'S FLOODED GARDEN

It was the winter after the steeply inclined grass field at the back of my plot had been ploughed over that my gardens flooded. The influx of water was so fast and prolonged that for days my vegetable patch sat under inches of water, the soil was waterlogged, and the crops rotted. The damage caused in such a short space of time was immense and, for me, it was a wake-up call. I needed to shore up my gardening defenses, and fast!

I understood how it had happened because I had written previously about soil and its structure with regard to climate change resilience. Once my neighbor's land had been turned over, its ability to absorb and retain water became greatly diminished almost overnight, and this was the devastating, gushing result.

To compensate, I created a swale and berm along the back of my garden, with a planting of trees, soft fruits, and fast-growing willow to help absorb some of the excess moisture. Allowing the grass to grow long also helped slow the flow, as did the addition of raised beds, interlinked with permeable gravel pathways. These pathways now allow me to walk on the ground without causing damage as the water slowly seeps away. Additionally, the use of much longer-lasting perennial plants and winter ground cover on the vegetable patch provide greater protection against the risk of valuable nutrients leaching out of my soil.

These measures, combined with the vital addition of well-composted soil, mean my gardens are now able to absorb (and stand firm against) a much greater volume of water than they would have been able to previously.

HEAT AND DROUGHT

Originally, when we first heard about climate change, it was in terms of increasingly Mediterranean-like summers for much of the temperate world. Images abounded of attractive hot, balmy evenings and the rather exciting potential to grow a wider range of produce outside and participate in lots of enjoyable alfresco dining on our homegrown fare. We remember hearing people joke at the time about how good climate change sounded. Many were actively looking forward to the prospect of this projected scenario we'd all been sold. Yet, the reality isn't as clear-cut by any means.

As we've already explored, the climatic changes and overall temperature increase will indeed herald more heat waves, it's just that it will occur alongside more extreme weather patterns in general. So, the picture we were painted was just a snapshot, a rose-tinted favorable segment, if you will, of the weather that we now face.

It's probably no surprise, given the previous projection, then, that scorching summers, huge forest fires, and widespread flooding have really flipped the switch on the reality that change is indeed upon us. The multifarious winter storms and flooding obviously didn't fit so clearly into this cozy, ingrained image of what global warming actually is. Nonetheless, prolonged periods of heat during the summer months (while not a given) are an increasing probability. Although sunshine is preferable to rain, many weeks of it (and soaring temperatures to boot) can be very problematic indeed for the garden, as many of us have seen in recent years.

LEFT: The telltale signs of drought.

As well as the time-consuming burden of dealing with watering to try and keep plants suitably quenched, the heat stress of such prolonged, high temperatures, especially for areas of the garden in direct sunlight, can cause immediate damage in the short term, stunting or stopping growth, or in some cases causing plants to wilt away altogether and die.

In the longer term, the stress of dealing with the heat can also weaken plantings that have otherwise (to all outward appearances) managed to stand firm. In such cases, it can hamper their ability to cope with further extremes, such as those potentially thrown at us in winter.

Protecting Plants from Heat Waves

Alongside reconsidering and revising the planting that we actually use in the future and growing for resilience moving forward, there are also quick and easy measures we can take to provide first step protection and prepare for the next time we experience such searing tropical heat.

Water at the Right Time of Day

A soaking either early in the morning or later at night will enable water to permeate much deeper into the soil to the benefit of your plants. Once the thermostat begins to rise, plants will start transpiring, and some of your precious resource, and the time spent delivering it, will simply evaporate away.

Water for Longer, Less Often

Rather than a surface watering, go for a longer soaking, so that the water seeps deep into the soil and stays around for longer. When you consider that the soil's surface forms the front line, enduring the heat of the sun, it makes sense to ensure your plants have access to water deep in the ground around their roots. This way, less day-to-day watering will be necessary.

Mulch Thirstier Plants

Not all plants have the same requirements, so those with greater thirst can be protected by way of a surface mulch around their base. Compost enables soil to hold on to a greater volume of water than it would be able to otherwise, so it's a first choice in this regard for a mulch, although other materials can also be used (see page 70). If you water before applying the mulch, the moisture will stay in the ground for even longer, protected by the layer above.

Vertical Watering

You can improve the infiltration of water through the soil by pushing the stiff stems of sweet corn or sunflower deep into the ground to create a pathway for water to move freely. In the same way, you can use a tool to create a narrow vertical hole, which you can backfill with sand.

RIGHT: **Watering plants early or late in the day allows the water to soak deeper into the soil.**

CASE STUDY

KIM ON GARDENING WITH A VERY LIMITED WATER SUPPLY

Like many homesteaders, my water comes from a private well and not the main supply. While this flummoxed me more than a little when I first moved from the city eleven years ago, I was assured (and shown evidence) that the supply of water on the land was solid and viable to support the house and gardens year-round.

That it did for many years, until the heat wave of 2018, when my well promptly dried out almost entirely. I could see it happening slowly but surely as the water tank took longer and longer to fill. Finally, we got down to a slow trickle, which was barely enough to flush the toilets in the house let alone anything else, and certainly not with any surplus for the gardens and polytunnels (hoop greenhouses) I have on my land.

Thankfully, with a community bartering spirit in good supply in my neck of the Welsh woods, my farming neighbors agreed to bring water to a tank every few days in exchange for a supply of fruit and vegetables from the garden. Not wanting to milk this goodwill to the detriment of neighborly relations, combined with the fact that accessing this water supply involved a lot of to-ing and fro-ing with buckets, I decided that my gardens simply had to make do with very little in the way of refreshments. Instead, I saw it as a way of testing my climate change gardening skills to the limit. The polytunnels had to survive with a watering just once a week, while the vegetable patch, shrubs, and other plants were only watered a handful of times over the summer months.

Although some produce struggled with the severe shortage of water, especially cucumbers and fennel, most thrived surprisingly well because of the measures I had in place to make the best use of the water. I've never watered so little before! While gardeners all around me were bemoaning the amount of work involved in keeping their gardens quenched, I was able to use the experience to assess firsthand how resilient a lot of plants and crops can actually be with just a little care and attention along the way.

RIGHT: **Kim's garden coped well with little water.**

Building Longer Term Resilience Against Heat Waves

Here are some easy ways to help build drought resilience in your garden.

Choose Your Plants Wisely

Some plants simply have greater resilience and are especially drought-hardy. Trees, shrubs, and long-lived perennials tend to have deeper roots, which enable them to seek out moisture from a much wider area. You can read more about this topic in later chapters.

Cover the Ground

During the summer months, it's important to provide ground cover around more vulnerable plants. This really does become extra valuable during a heat wave. Protecting the soil from the sun with foliage and/or mulch helps maintain the moisture levels to the benefit of surrounding plants (you can read more on mulching in chapter 4). Also worth considering are other forms of ground cover, including bark and wood chips. Even gravel gardens have their place when it comes to keeping moisture where you need it most. (See chapter 4 and also A Gravel Garden on page 189.)

LEFT: **Calendula creates ground cover among vegetables.**

Gravel acts as a mulch too.

Scrambling nasturtium covers the ground.

A trellis of hops shades a path.

Taller crops on the vegetable plot will shade others.

Use Plants as Shelter

A south-facing garden will be in the front line when it comes to the impact of intense sunshine. We don't want to block out the light altogether, but creating some areas of partial shade can be useful. This can be done by carefully positioning taller plants, such as sunflowers, Jerusalem artichokes, or a trellis of climbing plants so that they cast shade over smaller plants. New trees can also be used to cast natural shade over the house (see chapter 10).

By ensuring your soil is of the best possible quality, its ability to hold and retain water is greatly enhanced. Not digging, tilling, or turning the soil further enables the beneficial community of bacteria and fungi to work their natural magic, providing a greater resilience to extremes of weather.

Water Harvesting

When rain is in ample supply, it's hard to imagine how precious this resource might become in the months ahead.

A typical suburban roof could collect as much as 6,340 gallons (24,000 L) of water in a year: that's enough to fill 150 water barrels, assuming you had the storage space! This brings home the sheer volume of this natural resource that is available for harvesting during the wetter months of the year.

Certainly, the contents of any standard-size water barrel run out quickly during a prolonged period of drought, and not many gardens have room above ground for more than one or two of these collection vessels.

Making the most of gray water from general household use is another viable option. From wash water to bathwater, the amount of this precious resource that is used up and flushed away is undoubtedly rather staggering, were we to measure it. Yet a simple pump can be used effectively to channel gray water into your garden, where it can be put to a valuable second use. Be sure to check local regulations before setting up a gray water collection and reuse system, as many regional governments have rules that must be followed.

A simple yet effective setup to harvest water uses an old roofing sheet, guttering, and some ubiquitous blue barrels.

Harvesting water from a greenhouse roof means there is a handy store of water nearby.

CASE STUDY

SALLY'S POND

Some years ago, when we were constructing my raised bed vegetable garden, I decided to build a water storage pond. I'd seen them in traditional gardens and thought them a useful addition. The pond was lined with fiberglass to give it an extra-long life. We took a downspout from a nearby barn and created an overflow pipe that goes into a natural pond outside of the vegetable garden.

It proved its worth in 2018, when the United Kingdom experienced a long summer drought. The pond was full at the start of the year and the water lasted the whole season, getting down to the last 4 inches (10 cm) just before the drought broke. Another feature of a pond is the ease of dipping a bucket or watering can into the water and to have it fill instantly—no more waiting to fill it up, saving valuable minutes when watering the planting beds. In addition, myriad visitors are seen at the pond, including birds and dragonflies.

Planning for the Future

If you are moving into a newly built house and have some cash to spend on the garden, a wise investment would be an extensive water-harvesting system with an underground water storage tank. If you are lucky, you may even find that the builders have already integrated a rainwater-harvesting system into the property.

The problem with a standard rain barrel is that it holds up to around 66 gallons (300 L), so it will fill quickly after heavy rain. The only way to harvest more water is to have a collection of barrels. Even a large 220-gallon (1,000 L) tank will be quickly filled by a roof area of just 323 square feet (30 m2). There is also a risk of frost damage, as the water inside will freeze during a cold spell. An underground tank (cistern) can store as much as 2,200 gallons (10,000 L) of water, and sometimes more, so that's a considerable improvement. Being underground, it takes up no space in the garden and the water remains cool. It is fitted with a pump so that water can be used for irrigating the garden.

If you have the space or a new garden, future-proof your water supply by installing underground tanks.

Different Ways of Watering

Where possible, water the soil around your plants deeply. The quickest and most efficient way to do this is through a handheld hose nozzle, which allows you to direct the water away from foliage (where it will soon evaporate) and to soak into the soil, where it will stay around for longer. Watering cans also enable you to control the direction of the flow but are very time-consuming to use.

A handheld hose nozzle is a quick way to water vegetable beds.

Sprinkler systems are often used in polytunnels and for lawns, and enable you to flick a switch, turn a tap, and go. They also enable you to accurately measure the amount of time spent watering, so it can be adjusted according to the requirements of the day. They aren't the most efficient systems, however, in terms of water use, as everything gets a soak in the vicinity, including pathways and foliage, but they are quick and easy to work with. In the future, as the cost of water rises, you might want to look for more efficient options.

Slow seeping systems, like drip irrigation or soaker hoses, are popular and can work well during the summer months, especially for water-hungry crops and plants in large pots. The slow drip ensures a sufficient supply of water during warm times of year. These systems are also easy to connect to a water barrel or spigot and can be done often without the need for the purchase of an additional pump. However, the slowness of the drip throughout the day will mean that some of the water will evaporate and be wasted during a heat wave.

LEARNING FROM THE PAST:
Burying Pots

Water is bound to become more expensive and with the likelihood of irrigation bans being more commonplace, we need to look at different approaches. One efficient and ancient method of irrigation, used by small-scale subsistence farmers in the drier parts of the world, uses buried, unglazed clay pots, called ollas, which are filled with water. These have a short but wide neck and a wide body, which makes them look like a bean pot. Being porous, water passes through the clay and out of the pot when the surrounding soil is dry, so the rate of irrigation is dependent on the plant's water use. It's a highly efficient method compared with drip irrigation and surface irrigation and is well suited to use in the greenhouse or polytunnel. You can buy these pots online now, but you could adapt a clay pot by sealing the drainage holes and using a clay saucer as a cover.

This system of buried pots doesn't need a pressurized water supply, is less likely to be damaged, and the pots don't need to be filled every day, just once a week or so. It's targeted irrigation, so you are not watering the ground. To install, dig a hole that's wider and deeper than the olla, fork soil around the hole to aid drainage, and mix in one-third compost or manure, plus some sand or grit to improve drainage. Place the pot in the hole so its rim lies just above surface. Gently firm the soil in place around it, fill with water, and put on the cover. The pot spacing depends on the crops. Place ollas 3 to 5 feet (1 to 1.5 m) apart for sweet corn and other tall plants, and 3 feet (1 m) apart for tomatoes, peppers, and other garden crops. Water regularly and don't let the pot completely empty, as this will slow down the delivery of water when you refill.

Pop a lid on the pot to keep animals from getting in.

This system works best with well-established transplants, as their roots have a better chance of drawing the water. It's also wise to set up the system early in the season before the natural water reserves in the soil have been depleted and when it can most help plants get established. Add mulch to reduce evaporation further. This system works well for most crops, especially spreading plants that create ground cover. Very thirsty plants, such as spinach and squash, may not get enough water on hot days. Also, check that it's not delivering too much water during wet spells, creating excess moisture that favors disease.

WIND, FROST, AND SNOW

Even if you don't live up high in an exposed location prone to strong winds and a greater battering from the elements in general, the increase in the number of storms and potential for frost and snow later (or indeed earlier) in the season means these are prospective weather events to be very mindful of indeed.

Not Wild About Wind

It's referred to in some circles as the gardener's foe because wind can cause so much damage to plantings in a short period of time. At lower speeds, wind performs a valuable service, blowing the cobwebs (and indeed fungal diseases) away, so to speak. Light winds are welcome for good reason. Enabling sufficient airflow provides good ventilation within a polytunnel or greenhouse, which is essential to the health and vitality of indoor-grown plants. Outside, of course, this happens naturally; it's just that sometimes wind blows too often and too vigorously.

Even on a warm spring day, the temperature around your plantings can be significantly colder due to the cooling effect of the wind. In addition to potentially stunting the growth of seedlings early in year due to this temperature dip, wind can have a drying effect on foliage (known as wind scorch). It results in the browning of leaves, especially those of evergreens, which are most susceptible to wind damage. It can cause branches and foliage to rip and tear, and in extreme situations, wind can result in trees being uprooted and overturned. This can also cause damage to surrounding plantings and occasionally people's homes.

LEFT: **Frosted Swiss chard.**

There are many things that can be done, however, to provide a degree of protection for your garden against excessively strong winds, even in the most exposed of locations.

Planting Windbreaks

Plant trees or bushes as a windbreak, rather than building a solid barrier. This will take the sting out of the gale by slowing it down. A complete barrier would reflect the wind away, in another direction, potentially causing damage elsewhere, which is why walls and fencing make less desirable windbreaks. Plantings provide natural protection against the wind and have the additional benefit of helping to soak up excess rainfall through their deep, wide-ranging roots, which also help bind the surrounding soil together. Just some of the suitable species in this regard include:

- European beech (*Fagus sylvatica*)
- Holly (*Ilex* spp.)
- Laurel (*Laurus nobilis*)
- Leyland cypress (*Cupressus* × *leylandii*)
- Common hornbeam (*Carpinus betulus*)
- Willow (*Salix* spp.)

Strong winds cause loss of roadside trees.

TIP

The sheltering effect of a hedge can be felt some distance downwind. Wind speed is reduced by 75 percent at a distance of ten times the height of the hedge. So, a 16½-foot (5 m) high hedge will provide shelter for at least 164 feet (50 m).

CASE STUDY

KIM'S WINDPROOFED GARDEN

My garden lies in a very exposed location, more than 650 feet (200 m) above sea level. Westerly winds, in particular, can be problematic. To provide protection, an outer layer of mixed trees have been planted at the back of the site, with a range of fruit trees located to provide a further natural barrier in front. This is topped off with fast-growing willow at the head of the vegetable patch. This multilayered defense ensures that the produce and plants in the central gardens are much better able to ride out storms.

Jerusalem artichoke provides a useful summer windbreak.

A row of crab apples creates an attractive windbreak.

Annual Windbreaks

In the vegetable garden, taller produce such as artichoke thistle, sunflowers, Jerusalem artichokes, and even runner beans can be used to provide a degree of wind protection to the other plants there.

Ornamental Windbreaks

The likes of willow, bamboo, and grasses can form an elegant, ornamental windbreak in the garden, and they can look extremely attractive as they are blown hither and thither by the wind. Another option to consider are fruit bushes and trees to provide both functional and edible protection.

Other Methods of Wind Protection

There are some other useful options in the battle against the wind. A trellis, with its spaced lattices, works to slow—rather than deflect—the wind, while an archway or a pergola populated with climbing plants can provide another aesthetically pleasing defense.

More drastic measures include the creation of a windbreak bank, which is essentially a mound of earth to deflect wind up and over your garden. It can be planted with hardy perennials to make an attractive and functional garden feature.

It's Snow Joke

Let's face it, a snow-covered garden looks pretty, and this icy material can actually work to provide a protective insulation to the ground and the plants buried beneath it. Think of an igloo to understand how snow can offer protection to plants. And of course, in winter, plants are generally prepared for the cold, having died back or gone into dormancy.

However, changes in weather patterns and the increased volatility that goes with them now mean that snow can arrive much later in the year—well into spring, in fact. This late uninvited arrival can cause a lot of damage. At this time of year, plants are beginning to grow new shoots and blossom, and many of these delicate structures simply can't tolerate extremes of cold. New growth will be damaged, weakening the plant in the process, and even killing more cold-sensitive species.

Storm SOS Checklist

If you've heard that a severe storm is on the way and you are worried that it will cause damage to your garden, here are some actions you can take to shore up your defenses:

- Move any pots with tall plants indoors or to a more sheltered spot in your garden.
- Protect taller produce in the vegetable patch (such as sweet corn and sunflowers) by adding some bamboo canes for staked and tied support.
- If you have any top-heavy plants, consider trimming them back to reduce their wind resistance and help them ride out the storm.
- Move any lightweight garden furniture into a garage or other secure place.
- If you have a polytunnel, make sure the doors are closed securely and cannot blow open. Also, check that there are no objects lying around outside the tunnel that could be picked up by the wind and damage the plastic. Make sure that everything is neat and tidy inside, as the wind can pick up objects inside as well as out.

An urban garden during a snowstorm.

Here's what you can do to help:

- Prune less on tender plants to discourage new early growth.
- Let plants stand through the winter. Old foliage can provide a degree of protection for the crown of a plant and help it weather a cold snap (as well as provide a potential overwintering habitat for wildlife).
- Choose your planting areas wisely. The lowest area of your garden will be the most susceptible to becoming a frost pocket, so bear this in mind when deciding where to plant in the first place. It's best to avoid this location altogether for frost-sensitive plants.
- Protect your pots, whether by moving them indoors or choosing frost-proof materials to prevent cracking or splitting when the cold weather strikes. It's best to be prepared in advance.
- In the case of heavy snow, the weight can cause damage to the branches and leaves of trees and shrubs. It's best to carefully brush them off to avoid damage.

Snow on a polytunnel or greenhouse should be removed to prevent tearing or cracking of the overburdened cover materials.

Risk from Frost

As the climate warms, it's likely that we will experience fewer frosty mornings, but there is always the risk of unusual weather patterns bringing a particularly early or late frost. Frosts form on clear, cold nights when the air temperature falls and excess water vapor in the air condenses to form dew on the surface of plant leaves. If the temperature drops below freezing, the dew freezes to form a frost, with the cooling continuing when the sun rises again in the morning.

What's a Frost Pocket?

Sometimes gardeners tut-tut about frost pockets and frost hollows, but what are they?

A frost pocket or hollow is basically a piece of land that has a higher risk of suffering from frost and is therefore extra prone to early or late frosts, both of which are bad news if you are growing sensitive plants in your garden. When temperatures fall at night, the cold air sinks to the ground and flows downhill to collect in hollows in the landscape. These low spots enable the cold air to hang around for longer. Sometimes the cold air may be trapped by a hedge, fence, or wall. If you know you have a frost pocket, you can help by making gaps in the barrier at the lowest point so the cold air can continue to flow away. Or you can prevent the cold air from flowing into your garden in the first place by putting up a line of shrubs or trees above the frost pocket.

Most brassicas are frost hardy.

This dahlia was damaged by an early frost.

Frost Damage

The damage caused by frost varies according to the hardiness of the plant concerned. For example, a subtropical plant won't tolerate any frost, but a hardy, winter-resilient species can cope with weeks of freezing weather. A hardy plant can survive prolonged freezing temperatures either because it is dormant in the ground or because the freezing point of its cells is much lower.

Ultimately, each garden has its own microclimate, so temperatures will vary from one spot to the next. If you garden on a slope, the bottom of the incline is likely to be more prone to frost than the top. This is, therefore, an important determining factor when it comes to deciding where in the garden to position less hardy plants.

Kales, such as cavolo nero, are tolerant of frost.

The physical damage from frost is caused when the water within cells freezes. When water freezes, the liquid turns to ice crystals and it expands by approximately 9 percent. It's this expansion that ruptures the plant cell walls, damaging the plant in the process. Plants are particularly vulnerable to late spring frosts, because, if they get early morning sun, this can warm up the plant too rapidly, which can also damage the leaves. Should you find a plant has been frosted, you can spray it with cold water before the sun reaches it and this will help reduce the degree of harm caused. Try not to cut back perennials before winter, as the old shoots provide protection for the plant until spring. For the same reason, it's best to wait until the risk of frost has passed before pruning shrubs in the spring.

If you know a late frost is forecast, you can:

- Cover small fruit trees with a row cover at night to protect the blossoms.
- Wrap vulnerable container plants with a row cover.
- Cover small plants with a cloche.
- Well-watered soil holds more heat than dry soil does, so water a seed bed or young transplants early in the day and then cover them with a row cover.
- Avoid hoeing on a day that frost is forecast, as hoeing has been found to lower the surface temperature of the soil overnight by several degrees.

How to Save Damaged Plants

Is there anything that can be done to save a frost-blackened plant in the garden? Well, if it's a soft-stemmed plant, remove the frost-damaged shoots, as they may rot and cause more problems. For a woody plant, leave the frosted shoots alone until spring, when the last risk of frost has passed. Then prune out all the dead stems. The living stem should start to grow back in time. As with plants that have suffered from flood damage, encourage new growth with a fertilizer boost.

HEALTHY SOIL

The soil in our gardens may not be glamorous or indeed exciting to look at. It's easy to take for granted the stuff in which we dig holes to plant. Yet it's so incredibly important when it comes to building resilience against climate change. A healthy, nutrient-rich soil is at the heart of matters. It should be at the forefront of our efforts to provide protection against the increasing extremes of weather that face us.

As well as helping to mop up carbon dioxide and forge stronger, healthier plants, soil is a living entity. There is a complex web of interdependent microorganisms living within it and they need to be looked after. There is a vital connection between soil, plants, animals, people, and the planet. For example, healthy, compost-rich soil has an improved structure that enables it to absorb and retain more water than it would be able to otherwise. That's very handy when you consider that our precious soil has to stand firm against the future weather, which could include deluges of rain alongside periods of searing heat and drought.

With higher summer temperatures in the future, the soil will be warmer and there will be more evaporation of water from the soil's surface and plant leaves. Plant roots will need to take up more water to replenish the losses. In fact, a rise of 5.4°F (3°C) by 2080 could see the average soil moisture content reduced by as much as a quarter. So, a healthy soil is at the top of the list of things we should be aiming for in our gardens.

How can you achieve a healthy soil? We are going to guide you through the components that make up soil and the life within it, and outline simple yet effective ways to nurture and improve your soil to the benefit of your climate change garden.

LEFT: Healthy soil leads to healthy plants, such as these beets.

What's in Soil?

Soil is a mix of minerals, organic matter, air, water, and living organisms. The minerals—sand, silt, and clay—make up almost half of soil and they come from the breakdown of rocks. The organic matter comes from the breakdown of dead and decaying matter, while the spaces between all the particles are filled by air or water. The ability of soil to retain water depends largely on its texture. Soils containing more of the smaller clay and silt particles can hold much more water than a sandy soil with much larger particles.

Why Is Organic Matter Important?

Soils vary in their organic matter content. Organic matter is important because it forms a reservoir of nutrients, so the amount of organic matter present in soil is an indicator of fertility. The organic matter helps soil particles clump together and form aggregates, and this improves the structure of the soil, which in turn improves permeability and the ability of soil to hold water. Organic matter acts like a sponge, holding up to ten times its weight in water, which can be used by plants.

The Vital Component

Living organisms make up less than 1 percent of soil volume, but they are a vital part. They range in size from bacteria to earthworms. Bacteria and fungi are key to the decomposition of organic materials, while specialist bacteria take on other roles. The nitrogen fixers, for example, take nitrogen from the air and convert it into nitrates that plants can use. Some nitrogen-fixing bacteria are free-living in the soil, while others are found in the root nodules of leguminous plants, such as peas and beans.

Also important are the mycorrhizal fungi that are associated with plant roots. A mycorrhizal fungus supplies a plant with nutrients, while the plant supplies the fungus with sugar. Mycorrhizal fungi have a very large network of thread-like hyphae that extend out from the plant roots, enabling the fungus to source nutrients and water from a much larger volume of soil than the plant could do itself.

Climate change will affect soil life too. As temperatures increase, biological activity will increase. Soil microbes will be active for longer as spring arrives ever earlier and winters are milder. This may lead to more rapid breakdown of dead and decaying materials, so more nutrients may be available to plants.

Quick Healthy Soil Check—Do You Have Earthworms?

The presence of earthworms is a great indicator of soil life and health. Earthworm numbers decline if the soil is compacted, waterlogged, too acidic, or has been tilled or turned, but their numbers increase if there is plenty of organic matter in the soil. The best time of year to count earthworms is in early autumn or late spring, times when earthworms are active and can be found in the upper layers of soil. It helps to carry out the count after warm, wet conditions too.

Dig a soil pit about 8 × 8 × 4 inches (20 × 20 × 10 cm) and place the excavated soil on a tray or plastic bag. Break up the soil and collect all the earthworms. Count the total number plus the number of adults and juveniles. Size doesn't tell you about maturity, as earthworm species vary in length. An adult earthworm is identified by the presence of a saddle, which is a thickened ring of segments about a third of its length from the head (pointed end). A healthy soil will have ten to fifteen earthworms in the block. Once counted, put all the earthworms back in the hole with the soil.

It can be useful to look in more detail at the types of earthworms in your garden. There are many different species ranging in length and thickness, each with a specific role. You can download earthworm identification sheets from the internet and see what earthworms are in your garden.

Examine Your Soil

Before you can improve your soil, you need to know what type it is and whether there are any issues, such as compaction. Here are a few basic tests you can complete to determine your soil type.

Soil Texture

Take a handful of soil and squeeze it. What does it feel like? Sandy soil will feel gritty and won't hold its shape, whereas a handful of clay-rich soil will feel wetter, slippery, and will hold its shape. If it's something in between, then you have loam—a mix of all three particle sizes. The ribbon test is easy too. Roll some damp soil between your fingers and thumb. If you can squeeze it into a ribbon that's 2½ inches (6 cm) long or more, it's rich in clay. A shorter ribbon indicates a loam, while soil that won't form a ribbon at all is rich in sand.

Take a good look at your soil.

Slake Test

Slaking is the breakdown of large soil aggregates into smaller aggregates when suddenly immersed in water. This test examines the soil's crumb structure. The slower the soil breaks up, the more organic matter present. Take a handful of soil and place it in a bag. Fill a shallow dish with rainwater and place on the side. Select three pea-size lumps of soil and, using a teaspoon, gently place the lumps in the water, spaced equidistant. After two hours record how much the lumps have broken up and then again after twenty-four hours. Soils with aggregates that hold together have more organic matter, which helps bind the soil particles together. They are better able to maintain their structure during wet weather. These soils are more resistant to erosion and can maintain their structure, providing adequate air and water for plants.

Percolation Test

The percolation test looks at drainage. Dig a square hole about 12 inches (30 cm) across and 12 inches (30 cm) deep. Fill the hole with water and allow it to drain away. Then fill it again and record. Record how long it takes for the water to drain. Ideally, you are looking for a soil that allows water to drain away in two to four hours; anything outside of this range means you need to take some remedial action. If drainage times are very slow, even as much as ten hours, you may have compaction and/or a hardpan under the topsoil is stopping the water from draining away.

A slow-draining soil is liable to stay waterlogged, and this can lead to the death of plant roots. In contrast, if the water drains in an hour or less, the soil is very sandy. Any water will drain away from the plant roots too quickly and in dry weather the plants will need a lot of watering.

Biochar

Sally is fascinated by biochar and the potential it offers to the climate change garden. Biochar has long been made in the Amazon by indigenous tribes and used to improve the productivity of their soil. Plant matter is placed in trenches, covered with soil, and burned slowly in low-oxygen conditions to make a charcoal. Today, biochar is sold as a soil amendment. It's made in the same way, from plant biomass by pyrolysis (that's burning at high temperatures without oxygen) to create a black material that is rich in carbon. It looks solid, but it's filled with lots of microscopic holes, so it's porous and rigid and has a huge surface area.

Mixing biochar into soil has many useful benefits: it improves aeration, reduces the risk of compaction, and holds on to soluble mineral nutrients so there is no leaching. Soil microbes shelter in the particles and hide from larger predators, so the number of beneficial fungi and bacteria within can be ten times more than in soil without biochar. One property that is important for a climate change garden is biochar's ability to hold on to water. It seeps through the honeycomb of holes and does not drain away, boosting the moisture-holding capacity of the ground. Research shows that the use of a biochar-amended soil around transplanted saplings leads to greater survival rates because it improves the plants' ability to withstand drought.

Biochar can also help efforts to counter climate change because it locks up carbon, so there is less carbon dioxide released into the atmosphere when compared to burning plant biomass or leaving it to rot. In fact, the formation of 1 ton of biochar from wood removes 3 tons of carbon from the atmosphere.

How much biochar to add? It's almost a case of the more the better. Some people say a thin layer on top of the soil is a good start. Another way is to mix biochar into your compost as you add it as a surface dressing. Whether dug in or left on top, biochar will boost soil health. You can mix it into the potting mix used for potted plants at a rate of 4 ounces (100 g) per gallon (4 L) of compost or into raised beds. Permaculturalists say it's important to "biocharge" the biochar to get accelerated growth of mycorrhizal fungi by mixing biochar into your compost heap so that it's full of microorganisms before it goes on the soil.

LEARNING FROM THE PAST:
Dry Farming

Dry farming is farming with little or no supplemental water and is practiced in the western United States, Mexico, Mediterranean countries, on volcanic islands such as Lanzarote, in parts of Australia, and even by the Incas more than 11,000 years ago. We can learn so much from these methods and apply them to our own gardens.

The formal definition of dry farming is growing crops during a dry season using the residual moisture in the soil accumulated during the wetter parts of the year and with no additional irrigation. It's not a method to maximize production, but rather it's a method to maximize sustainability because yields are much lower. Grapes are the traditional dry-farmed crop, as their roots extend deep into the ground in search of water and minerals. Dry farmers have various strategies to conserve soil moisture, including the incorporation of organic matter, no tillage/no dig, and mulching, combined with the use of drought-resistant varieties and carefully timed planting. Key to success is soil type and water-holding capacity, because it's not possible to use these methods on well-drained sandy soils with little organic matter.

Native Americans of the desert states of the United States, such as the Hopi, have been farmers for thousands of years. They cope with intermittent winter rains and grow drought-resistant crops, such as maize (Hopi blue corn). They harvest water during storms by building a series of canals to direct the water to shallow beds, where crops are grown.

On the volcanic Canary Islands, farmers build small walls of lava rocks around their beds to create shelter and trap condensation from fog. Similarly, West African growers build low earth berms around key plants to trap water and provide shade from sun and wind, which helps seedlings to get established.

Well-Draining Soil

We'll be using the term "well-draining soil" a lot. It describes a soil that has a good structure with lots of air spaces that allow water to percolate through at a reasonable speed, neither too quickly nor too slowly.

Drainage is one of the most important factors in making your garden climate change resilient. It is difficult to tell just from looking at soil how well draining it is, but if your soil is always soggy and puddles lie around for several hours after rain, you've got poor-draining soil (see the percolation test on page 66).

The soil type in your garden determines which types of plants you will be able to grow and gives you an indication of the problems you might encounter during extreme weather. Sandy soils are great when it rains heavily, as the water drains quickly, but in a hot and dry summer, these soils are going to dry out more quickly and require more watering or mulching. Clay soils hold water in winter and are prone to waterlogging, while in summer they dry up and become hard and difficult to work, but they do hold water for longer.

Organic matter is usually the answer to better drainage. Adding organic matter, such as compost, to a sandy soil will improve its structure and allow more water to be retained. Similarly, mixing organic matter into a heavy clay soil will also improve its structure, helping water move through the soil more quickly and reducing drainage times. If you have particularly heavy clay, the soil may benefit from grit too.

A typical wedge of soil.

Mulch Materials

Something else that's recommended for improving your garden's resilience is mulching—spreading of a layer of loose material over the surface of the soil. The benefits of mulching are many:

- It helps retain moisture in summer.
- It suppresses weeds.
- It shelters plant roots from excessive high temperatures.
- It helps water penetrate the soil rather than running off the surface.

All sorts of materials can be used as a mulch, some inert and others that rot down. Organic materials that rot provide the additional benefits of adding organic matter, supplying nutrients, boosting drainage and water-holding capacity, and encouraging microorganisms. Biodegradable mulches include grass clippings, wood chips, sawdust, bark, pine needles, shredded leaves, leaf mold (composted leaves), ramial wood chips, wool fleece, manure, straw, hemp, compost, seaweed, and shredded paper.

Grass Clippings

Grass clippings are a low-fertility option and are spread thinly over the soil after watering or rain. Do bear in mind that too thick a layer will create a foul-smelling mat of rotting grass. Even a small application retains moisture and suppresses weeds. Some people allow the clippings to dry first before adding a thicker layer over the soil. It's important to always check first that the lawn from which the clipping have been taken hasn't been treated with chemicals such as weed killer or pesticides.

Wood Chips, Sawdust, Bark, Pine Needles

Some people argue that these materials rob the soil of nitrogen, but research has found that if they are mulched rather than dug in, they are fine to use. They quickly form a layer that suppresses weeds and reduces evaporation. In time, they will rot down and boost the organic matter. Be careful with sawdust, as it has a tendency to create a mat, but it can be used to create a water-permeable walkway. If you are using conifer wood chips, take care not to mulch around young plants and just use it around established plants because the tannins in the conifer wood are phytotoxic and can inhibit the growth of young plants. Also, if you are using chipped materials supplied by an arborist, it's important to ensure you do not use woody materials from plants suffering from bacterial canker or honey fungus.

Shredded Leaves

This is a good source of mulch, spread thinly over the soil surface.

Leaf Mold

Place leaves in a 1-ton bulk bag or in a wire-netting compost bin and leave them until they have rotted down to create leaf mold, but be aware that this can be a slow process, taking as long as two years. Leaf mold is not as rich in nutrients as compost, which is why it's sometimes used to make potting mixes. It's also a great stimulant for soil microorganisms.

Cucamelons with wool fleece mulch.

Ramial Wood Chips

These are made from young branches no larger than 2¾ inches (7 cm) in diameter. This wood has more nutrients and less lignin that older wood, so it's easier and quicker for fungal decomposers to break it down. Michael Phillips, author of *The Holistic Orchard*, has found that ramial wood chip mulch can lead to healthier trees.

Wool Fleece

A layer of fleece spread over the soil will reduce water loss and suppress weeds. It is claimed to be a good barrier for slugs and snails because they don't like moving over it. A layer of wool can be used to reduce water loss from pots in summer and provide insulation in winter.

A biodegradable plastic film acts as a mulch.

Inert Materials

There is a range of inert materials that can be spread over soil and used in pots, including gravel, pebbles, glass chips, slate, crushed shells, plastic sheeting, and landscape fabric, among others.

Ornamental mulch of recycled glass chips.

When to Mulch

You can apply mulch all year-round to avoid leaving soil bare as much as possible. Soil is more vulnerable to the elements when left exposed. However, the optimum time for mulching the vegetable garden is autumn, when you can spread the mulch after harvesting your crops. Mulch keeps the soil covered during winter and stops weed seeds from germinating, plus microorganisms have plenty of time to get to work before you sow the spring crop.

Mulching around shrubs and trees will also suppress weeds and create a moisture-retentive layer, but don't spread the mulch right up against the trunks, as it can promote rot. You can add mulch any time from late spring through summer when needed. If the weather is dry, water the soil first and then spread your mulch.

LEARNING FROM THE PAST:
No-Dig History

The "no work, deep mulch" method of gardening goes back to Ruth Stout in the United States. Ruth Stout was born in Kansas in 1884 and much of her gardening was devoted to labor-saving methods, and her book titles reflect that: *How to Have a Green Thumb without an Aching Back* and *Gardening without Work*. Her "no work, deep mulch" method was to keep a thick mulch of 8 inches (20 cm) of biodegradable matter on the vegetable and flower beds all year-round. As it decays, the mulch enriches the soil, and she would continually add more. She claimed never to have used a plough, spade, or hoe, or to have cultivated, weeded, watered, sprayed, sowed a cover crop, or built a compost heap—so she really did break the rules.

All sorts of organic materials were used to mulch her beds, including hay, straw, leaves, pine needles, sawdust, and vegetable peelings. In fact, she used anything that would rot. She sowed directly into the mulch, moving the mulch back to drop in the seeds and then covering them over again. During the year, if she saw weeds popping through, she added a handful of hay. She mulched whenever needed. From reports of people using the method in the United States, it works well in the hottest of summers, requires little or no watering, and is good for growing vegetables in poor soils such as sand and heavy clay.

Planting into thick mulch.

Ruth Stout's deep mulch approach was the forerunner of the no-till or no-dig methods that are now very popular. The advantages of this method are many: the compost layer covers and protects the soil, smothers weeds, boosts organic content, and improves water retention. In time, the soil builds up more organic matter and has a better water relationship between plants and soil; the earthworms and network of fungal hyphae are undisturbed, as is the capillary structure of the soil. In terms of water and flooding, most of the gardeners at Sally's farm are no-dig, and it's interesting that those who like to dig, or worse still, rototill, have seen more problems with puddling and even flooding compared with their neighbors who apply lots of mulch instead.

No-Dig Benefits for Climate Change Resilience

Research is increasingly highlighting the many benefits of not disturbing the soil. For a start, soil holds in carbon, and as the documentary *Kiss the Ground* highlighted clearly, not digging the soil helps keep carbon in the ground, safely out of the way. Turning soil over releases carbon into the atmosphere, therefore fueling the argument that soil is essential in the battle against climate change. Recent research by H. V. Cooper and colleagues into till versus no till on farms further confirmed that turning soil releases carbon dioxide into the atmosphere (Cooper HV, Sjögersten S, Lark RM, et al., 2021). They also found that roots fare better in a no-dig system because they are better able to penetrate the soil, reaching further down to find water. Also, there was found to be less methane and nitrous oxide release when using no till. Overall, not digging produced 30 percent lower emissions of all three gases. Earthworms were also found to be far more present in untilled soil because they are left undisturbed. As well as fertilizing the soil, these beneficial creatures create vertical burrows that aid drainage and boost microbial activity to the benefit of the plants growing there.

When you further consider that soils have lost 50 to 70 percent of the carbon they once held (that represents 25 percent of all man-made greenhouse gas emissions), practices such as not tilling are more important than ever. Every garden matters in the battle against global warming.

Good use has been made of mulch and the no-dig method.

Another technique that works well is leaving crop roots in the ground; chop off the old plants at ground level and leave the roots in the soil to hold the soil particles in place. The other benefit of these techniques is that you are locking up more carbon than you are releasing, helping in the battle to reduce carbon emissions. It may be a tiny contribution, but if every gardener did this, the effect would be great.

Composting

One of the most useful things you can do to create a more resilient garden is to make your own compost. This crumbly, dark, earthy-smelling material is the perfect soil amendment, and it's nature's way of recycling all your garden waste. If you have a tiny garden and no space for a compost bin, buy compost from commercial composting operations. However, it is far better to make your own if you have the space to do so.

Gardeners tend to have their own tried and tested way of making compost. Some simply pile it up and let nature take over. Others have a far more scientific approach, mixing up different materials and regularly turning the heap. A well-made, aerated, and moist compost heap made in summer and turned regularly can convert all the organic matter into compost within a couple of months, but a typical garden compost heap that is gradually built up from layers of materials collected throughout the year will take much longer.

But regardless of which strategy you use, to get a compost heap off a good start, you need a mix of carbon- and nitrogen-rich materials. There's lots of carbon in brown materials, such as leaves, woody material, straw, newspaper, and cardboard, while materials rich in nitrogen are the green things, grass clippings, coffee grounds, fresh leaves, and waste from the kitchen. You need roughly half brown to half green to get the right balance. Too much green and you will be left with a smelly mush. Too much brown and composting will be very slow.

TIP

Allowing the bottom of your compost pile to touch the soil enables the beneficial creatures involved in the decomposition process to make their way in. Equally, rather than meticulously turning your pile, another option to help speed the process along is the application of a layer of microbial-rich material from your finished compost pile.

A basic compost bin can be as simple as four sides made from recycled pallets and a lid to keep out the rain, but you can also buy plastic bins cheaply. It's best to gather all your materials together and fill the bin in one go. Cover and leave for about a month. Then take it all out, give it a good mix, and return it to the bin. Three or four months later it should be brown and crumbly. Any larger pieces of plant material can be removed, and the rest used on the garden. If you are filling the bin as you garden, then the heap won't get as hot, so the process will take longer, but it will get there in the end.

Compost—the Best Soil Cure-All?

Kim certainly believes so. Rather than testing her soil, Kim works under the principle that the best way to improve soil is through the application of a layer of compost on top. If you have a healthy loamy soil, this just needs to be done once a year.

Get up close and personal with your precious soil by sticking your hand in and looking for earthworm activity. You can see firsthand how it is faring. If your soil needs more of a helping hand, a layer of cardboard, followed by an application of 2 inches (5 cm) of compost and topsoil will enable you to get planting right away. It really can be as simple as that!

Why Organic Improves Resilience

Both of us are organic gardeners, and we feel strongly that using organic methods will put us in a better position to help our gardens cope with climate change. Why? It all comes back to soil health.

Soil is an amazing ecosystem and the living component—the microorganisms, worms, and other animals—is essential to plant health. Albert Howard, one of the pioneers of the organic movement, understood the link between healthy soil, healthy people, and a healthy planet. His years of research into soil and composting led him to conclude that the lower the state of fertility of the soil, the greater the likelihood of pests and diseases.

When you use artificial fertilizer, you upset the balance in the soil. It's not that the fertilizer is actually killing the soil life—unless, of course, you have added so much that the levels are toxic. It's about balance. Many different types of microbes help break down organic matter in a complex food chain and release nutrients in a form that plant roots can take up. If you add readily available nutrients, such as nitrates, this bypasses the ecosystem and the fertilizer can be taken up by the plant roots straightaway. This might seem like a good thing, but it's disrupting the natural system. A good analogy is that an artificial fertilizer is fast food for a plant; it's a ready-to-go food. By applying a fertilizer, you put all the "artisan" microbes out of work and they start to disappear. The longer you continue to apply the fertilizer, the lower the diversity of artisan microbes in the soil. This means that you have to continue to supply fertilizer because the supply of naturally sourced nutrients is drying up. As a result, the plant won't perform quite so well without an artificial boost, so it's a catch-22. Leading soil life expert Elaine Ingham describes soil that lacks a diversity of microbial life as "dirt"—and that's what intensive farmers with their reliance on fertilizers are effectively working with.

The other problem with artificial fertilizer is that it is soluble. It's ready for the plants to use, but it is easily washed away. It just takes a heavy rainstorm, and nitrate fertilizer in the soil will be washed deeper beyond the plant root zone or even out onto paths and into waterways. And as we know, heavy rain is going to be more common in the future.

Obviously, the application of pesticide and weed killer is also going to adversely affect some soil microorganisms and upset the balance. Even glyphosate, the one chemical we were told was not persistent and could safely be used without harming life, is proving to be harmful. The ecological safety of this chemical—the way it reacts with living organisms and how it is degraded by them—has not really been assessed. The manufacturers claimed it would be inactivated quickly in the soil and degraded by microbes, but now we know it is far more persistent and adversely affects soil life. The initial safety studies were carried out on just glyphosate, but glyphosate-containing weed killers contain other substances, called adjuvants, to improve their effectiveness. Now scientists are finding that these adjuvants can be more toxic than pure glyphosate or they react with glyphosate in a way that was not predicted. The science is still muddy. Studies have found that glyphosate can lower earthworm activity and reduce the viability of their cocoons, leading to fewer juveniles. Other studies have found that glyphosate has no effect on bacteria but increases numbers of fungi and actinomycetes. Still other studies have found that repeated applications have long-term effects on the balance of microbes, with those sensitive to glyphosate disappearing. Research published in recent years has found that even gut bacteria in bees were harmed by glyphosate, making them more prone to infection.

Combining an organic approach with techniques such as no dig, incorporating organic matter, and providing soil cover help prevent nutrient and water loss. Leaving the soil undisturbed allows fungal hyphae to remain intact, holding the soil particles together and creating a network of paths through the soil. Having a high diversity of plants, wild areas, and ponds all help make your garden more resilient to pests and diseases.

DESIGN IDEAS

How we cope with increasingly volatile weather will be crucial moving forward, especially as we are faced with everything from acute rain and cold snaps to long periods of summer drought and soaring high temperatures. Throughout the world—as you have already read in this book and no doubt picked up on in the frequent climate change news in the media—it's going to be an ongoing problem. Building resilience is therefore key.

Resilient Garden Beds

Thankfully, building resilient garden beds doesn't necessarily involve a complete redesign of your current gardening space, as in reality there are lots of hardiness-boosting measures and features that can be incorporated into existing layouts. They include simple things, such as raised beds, mound beds, and sunken beds. As we outline over the coming pages, it's also incredibly useful to learn from our ancestors and look back to see how gardeners coped with big freezes, heat waves, and even the Little Ice Age.

A Simple Raised Bed

We mention raised beds a lot because they help make your garden resilient. Why? Because they improve soil drainage, which is very important for plants that can't cope with waterlogged ground over winter. Also, they can raise your precious planting out of the low-lying danger zone of flooding. The ground around your beds may become saturated, but unless it's very deep floodwater you're dealing with, your crops and their roots will be safely high up above, out of harm's reach. The soil in the bed also remains unaffected and there is no erosion of the nutrients within.

LEFT: **A modern design with permeable gravel paths, raised beds, and water features.**

Despite Kim's garden being flooded, the plant roots in these raised beds were safely above the water.

Another benefit of raised beds is that the soil tends to warm up more quickly in spring. Also, because you're working with a smaller volume of soil, it's easier to adjust conditions (such as pH) easily, allowing you to grow plants with specific requirements, such as an acidic or calcium-rich soil, for example.

One potential downside of raised beds, however, is that in hot weather the soil dries more quickly and the plants may suffer. Because of this, it's important to watch the moisture levels and water if necessary. That said, using ground cover in between planting and a mulch around especially water-hungry plants keeps moisture where it is needed most and reduces watering requirements overall.

Raised beds made from recycled timbers.

Raised beds are typically used for vegetables but also work well for most other plants. The materials from which the beds are constructed can be pretty varied—anything from stone and brick to wooden planks and corrugated iron. Even old glass bottles. While a surrounding frame isn't essential for raised beds, having one offers many benefits, including the ability to provide further protection from whatever the weather decides to throw at you. These valuable structures are also very easy to build. Simply mark out the edge, clear any vegetation, check levels, mark corners, hammer in corner posts if these are being used, and secure the boards to them. You may need retaining stakes for the timbers and planks, and particularly deep walls may need footings. Then place a layer of thick cardboard over the soil or existing vegetation and backfill with soil enriched with organic matter. If the bed is in a low-lying area or over heavy clay soil, add 4 inches (10 cm) of gravel or crushed rock before backfilling. Then as a final touch, cover the paths around the raised beds with bark, gravel, or stone to create a permeable surface.

A mound bed is in the process of construction.

The wood chips have been covered with a thick layer of compost and planted with a mix of vegetables.

Pile It High for a Drought-Resistant Bed

Mound or hügelkultur beds, as the name suggests, are heaped beds. They are built by digging out a trench and backfilling it with a layer of woody materials, such as small logs, branches, and bark to provide a long-term carbon source that is then covered by more layers of wood chips and then a mix of soil and compost to create a mound that can be steep-sided or shallow. The numerous air spaces around the woody material help retain moisture and encourage plant roots to extend deep into the ground, creating a truly drought-resilient bed.

The conditions on one side of the mound compared with the other will be different; an east-west orientation means there is a sunny side and a shadier side, while the conditions at the bottom of the bed will be different from those at the top. This allows you to plant a range of different plants on the mound. Typically, a mound bed like this is used for growing vegetables, but a shallower version could be useful in the flower garden too.

Sally was impressed with her mound bed during a previous season's drought. It wasn't watered, other than a little for newly planted vegetables. The 'Queensland Blue' squash did exceptionally well compared with others grown nearby, producing 98 feet (30+ m) of shoot growth and twelve large fruits. Its deep roots were able to reach down to the moisture trapped by the layer of rotting woody matter. Another benefit of squash is that their large leaves shade the ground and reduce evaporation from the surface of the soil.

LEARNING FROM THE PAST:
Waffle Gardens

We can learn a lot from the dry-farming techniques of Native American Pueblo peoples, such as the Zuni from the Southwestern United States. For centuries, the Zuni have tended resilient gardens, making use of techniques that we see today, such as the three sisters method of planting corn, squash, and bean together; companion planting for pest control; and even no-dig methods. Traditionally, they used methods that avoided disturbing the ground—just moving the soil to create a hole in which to plant and leaving the rest alone—and using mulch to suppress weeds and retain moisture. Each household had a waffle garden, which was built close to the home and river for water, and these gardens are still used in Arizona and New Mexico today. They are the forerunner of the modern-day square-foot garden!

A waffle plot comprises rows of square cells separated by berms of compacted clay soil, forming a deep-walled container, which is effectively a sunken plot. When heavy rains come, it can't run off, but collects in the depressions and soaks into the ground. When looked at from above, the cells resemble the pattern created by a waffle iron. Vegetables are planted in the cells; the shape helps direct water to the plants while the berms throw shade onto the plants. In addition, the Zuni build a low wall around their gardens to reduce the drying effect of winds.

To make a waffle garden, you need a level site so the water flows over all of it, rather than to one side, and clay soil. Clear the ground and plan your squares, typically 1 by 1 foot (30 by 30 cm), but they can vary in size and shape to suit the site. Also leave space for pathways. When the soil is wet, make the cells using a hoe. Start in a corner and move the soil from the middle of the cell to the edge to build up four walls. Press the soil together to create a wall 5 to 6 inches (12 to 15 cm) high and 5 to 6 inches (12 to 15 cm) wide. Add water if the soil is too dry to mold. Once you have created a series of cells, backfill with compost but make sure there is plenty of room to add water without the water overflowing. Typically, the Zuni would use a range of soil amendments to create a fertile soil—flood plain sand, forest soil, sheep manure—and they may have mulched with gravel. And remember, you don't have to build up, you could dig down, creating a series of sunken pits, rather than raised berms.

When it comes to planting, make sure taller vegetables, such as corn, don't shade plants that like full sun. Similarly, place spreading vegetables, such as summer or winter squash, at the edge so they don't scramble over other crops.

Another Drought-Resistant Idea: Sunken Beds

Another option is to dig down. This may be a good alternative for gardens in particularly dry areas, as a sunken bed maximizes water collection and shelters plants from drying winds. Sunken beds are great for keeping plantings cool during a hot summer and help retain heat in a cold winter. Plus, they can be easily watered through flood irrigation. Sunken beds are traditional in the drier parts of the world, such as the Southwestern United States (see page 83).

From a practicality perspective, this type of bed does require a lot of effort, and indeed as you'd imagine, there's plenty of digging involved. On the plus side, though, you don't have to buy any materials to make one.

To get started, mark out the bed as usual and have your spade at the ready. Excavate a good depth of soil, around 24 inches (60 cm), although these beds can also be deeper. Put the first few inches of topsoil to one side but pile up the rest of the excavated soil around the edges of the pit to create berms (raised banks). Once it's been dug out, level the base of the bed, then refill with the topsoil mixed with plenty of organic matter. Fill to within 6 inches (15 cm) of the surface to allow space to mulch and water. Stomp on the berms to compress the soil so they are firm. The soil is less likely to be washed away in heavy rain. Flatten the tops so you can use them as paths. If you don't have enough compost, an alternative is to use the sunken bed as a compost heap for a year, slowly filling up the hole with organic matter until it's ready to be used.

LEFT: **The soil in this garden has been mulched while the gravel paths are permeable so water drains into the ground.**

A permeable slate chip path.

Slow Water

The "sink it, spread it, and slow it" approach to landscape design is just as applicable to your gardens as it is to cities. There are numerous ways you can slow down the flow of water across your outside space; these include permeable paths, front gardens, swales, rain gardens, bog gardens, green roofs, and living walls.

Permeable Paths

A permeable path may sound quite fancy, but it's not. It's simply a path that has been constructed using materials that allow water to pass through and into the ground below. Gravel, for example, is a freely permeable surface. It's cheap and easy to install. The path is dug out, leveled, and backfilled with stone chips and then topped with a layer of gravel or crushed slate. You could also use recycled materials, such as crushed glass. Another option is pavers or stones that are laid over a bed of sand and gravel so the water can pass through, or flagstones surrounded by gravel.

If the area is going to get a lot of foot traffic or be used as a driveway, you can use reinforcement grids, which are like a mesh. They are laid in the ground and the mesh spaces backfilled with gravel or soil and grass seed. Resin is another option for a large expanse, such as a parking area or large drive. Unlike concrete, a resin-bound driveway or path is permeable as well as being nonslip, resistant to weeds, and easy to maintain. Resin-bound materials comprise small stones bound together with a high-tech resin. It contains many small gaps, so water can drain away. It has the appearance of gravel but the durability and permeability of block paving. This is very different from resin-bonded surfacing in which the stones are scattered onto a surface of resin, which results in an impermeable layer.

A Front Garden

Next time you are walking along a typical urban street, have a look at the front gardens. Traditional flower beds and lawns have more often than not been replaced with hard surfaces, such as paving or a concrete pad for a car. The result of this surfacing, which is impervious to rain, is the considerable runoff we see from these urban spaces. Just imagine how much water could be slowed down if every resident of a street increased the permeability of their front yard.

It's also easily solved. Cars don't use the central part of the driveway, so it could be replaced with gravel or pavers, both of which allow water to soak away. Similarly, in the case of a sloped drive, a French drain or grate could be installed at the bottom to collect any water that runs off. Gravel trenches can also be dug along the edge of concrete and paved areas, again to slow down any runoff and let the water seep safely into the ground.

An attractive and functional front garden.

RIGHT: **The steel mesh creates a hard surface for a car and allows water to drain away.**

Water is directed down a slope to a rain garden.

Swales Large and Small

Swales are brilliant at slowing, spreading, and storing water. Comprised of a shallow trench that carries water safely away from areas at risk of flooding, swales particularly come into their own in gardens at risk of water flowing in from above. This simple measure can help distribute the water so it can drain slowly over the garden and into areas that need it.

Before you start digging, though, remember to watch how water moves through your garden. Find out where it is coming from and which route it takes. Also bear in mind that swales are suited to shallow slopes of 5 to 10 degrees or less, any more and you risk the chance of a mudslide during heavy rain. A swale needs to be positioned with the water draining away from any building, steep slope, or septic bed or drain. Ideally, a swale is located uphill from a point where water collects.

To build a swale, dig a trench along a contour line, up to18 inches (45 cm) deep and 24 inches (61 cm) wide. Its length depends on how much space you have and how much water needs to be stored. You can work out how much water your roof and hard surfaces collect, based on the annual rainfall for your area (see page 25 for how to calculate this). This allows you to see whether the capacity of your proposed swale is enough. When you dig the soil from the trench, pile it up on the downhill side to create a berm. Make sure the bottom of the trench is level so that the water lies evenly and doesn't accumulate at one end. Don't fill it immediately but leave the swale empty and watch during the next heavy rainfall.

If the water overflows, the trench needs to be deeper, longer, or wider. If a lack of space limits the size of the swale, you can always build a second channel to catch the overflowing water from the first. Once you

At the National Botanic Garden of Wales, this decorative granite paver channel in the path is snaking down a slope and carrying water away from the rest of the garden.

are happy with the swale, plant the berm with herbaceous perennials, fruit bushes, fruit trees, or even hedgerow plants. The roots of these plants will stabilize the berm and their leaves will protect the soil. The swale itself can be left empty or backfilled with gravel or mulch. And a final touch is to include an overspill area to deal with very heavy rain.

If you find your garden experiences heavy runoff and the swale can't cope, lay a porous pipe along the bottom of the channel, which will help move the water along more quickly and out through the overspill.

Don't have room for a large swale? You can build a series of mini swales, shallow trenches backfilled with absorbent material and mulched with bark. A series of mini swales and swale paths can direct water through your garden and allow it to soak into garden beds. Another advantage of water soaking into the ground is that you don't get muddy paths with pools of water sitting on the surface.

Rain Chains

Design in the garden can be fun as well as functional. Instead of downspouts, you can use rain chains to link a gutter with a drain or water channel (rill).

CASE STUDY

GARDENING IN THE SKY

Walking along the "garden in the sky" on New York's West Side is a surreal experience. You are high above the hustle and bustle of the busy city streets, enjoying a fabulous prairie planting with views over an iconic waterfront. This is the High Line, an inspirational park built on an abandoned elevated railway that was once used for transporting cattle into New York's meatpacking district. It fell into disuse and was about to be demolished before the Friends of the High Line took it over and transformed it into a beautiful public space. The High Line is a continuous 1.45-mile (2.3 km) long greenway with gardens, art, food stalls, and performance areas.

At the start of construction, everything was removed from the structure, including the rails, although many have since been put back in their original position, a drainage system was installed, and the structure waterproofed. Once the structure was safe, the landscaping began on soil that ranged in depth from 18 to 36 inches (45 to 91 cm). One of the landscape designers was Piet Oudolf, famous for his prairie planting style (see page 193). There are more than 500 plant species, including perennials, grasses, shrubs, and trees, all chosen for their tolerance of drought and wind hardiness, texture, and color, with a focus on native species. They include aster, birch, bur oak, catmint, chokecherry, dogwood, echinacea, geum, hydrangea, meadow sage, Michaelmas daisy, milkweed, salvia, sedum, smokebush, Virginia pine, winterberry, and yarrow.

Rain Gardens

A rain garden is just like a sunken bed, positioned carefully to collect runoff from surfaces in your outside space. In addition to temporarily holding stormwater and slowing down runoff from hard surfaces, it has the bonus function of acting as a natural filter for the water running through it. It's stocked with plants that can cope with being flooded on a temporary basis, thereby also making an attractive feature out of your useful ally in the battle against flood prevention.

The best place to build your rain garden is several feet (meters) away from the house on a flat area or gentle slope (less than 10 degrees). This way, swales and drainpipes can direct the water into this area, where it will collect and slowly drain into the ground (or if sheer volume of rainwater demands, into a further overspill area). As previously mentioned, this should always be situated well away from septic tanks and wells.

Also, do bear in mind that because you'll be relying on your rain garden to allow water to soak away into the ground, it's prudent to check that you've chosen the right place (and soil) for such a job. A simple percolation test (see page 66) is recommended before you start construction. You need soil that allows water to drain away at a rate of at least 2 inches (5 cm) per hour. If percolation rates are slow, you either need to find a different spot or improve the soil with gravel or grit. A rain garden may not suit gardens with heavy clay or a naturally high water table, as the water will be too slow in draining away. In these cases, a bog garden may be better suited to take advantage of the waterlogged conditions.

These valuable gardens can be any shape, but they do benefit from being as large as possible so they can effectively deal with an excess of rain in a heavy storm. With this in mind, aim for 10 feet (3 m) wide or more. As a general rule of thumb, the surface area of your rain garden should be at least 10 to 20 percent of the total area of impermeable surfaces that drain into it, including the roof. If it's too small, it will simply get waterlogged too quickly and overflow.

Construction phase.

A small rain garden has been constructed in this suburban garden.

To construct your rain garden, mark out the boundaries and dig out the soil to a depth of roughly 18 inches (45 cm), creating a shallow saucer-shaped depression. Make sure the edges are level and create an earth mound (berm) on the downhill side to hold the water with an overflow zone (which can simply be a notch in the berm that allows excess water to seep away toward a channel or drain). Mix organic matter and grit into the soil and refill the rain garden to the original level.

When it comes to planting, a good root system will hold the soil and take up water, and the more plants, the merrier. Also, in terms of construction, rain gardens tend to have three growing zones to bear in mind when choosing plants:

- The bottom of the rain garden will hold the most water and for the longest period.
- The sloping sides of the garden will be wet but for shorter periods of time. Ideally these plants need a good root system to stabilize the slopes.
- The upper area will be the least wet.

It's also good to use a dense and diverse mix of plants so that if some fail, others may thrive. Given the damp nature of the bed, avoid any of Mediterranean origin and go for species suited to wet or waterlogged conditions because they have to be able to cope with waterlogging around their roots on a temporary basis.

In the first year you may have to water in dry periods and while the plants are getting established. During this initial period, the plants will be less able to cope with loads of water, so don't allow too much rain to collect there by deepening the overspill area to let more water out. It may also help to place rocks near the entrance to the rain garden to slow down the flow of water and prevent the plants and soil from being washed out.

Another option is a rain garden planter—this is a large container that receives water from a downspout and has an overflow pipe leading to the drain. The planter intercepts the rain and holds it back before discharging to the drain, rather like a rain barrel, but far more attractive. The planter has a deep gravel layer at the base and is then filled with compost mixed with grit and planted with water-loving plants such as iris, sedges, and rush.

Bog Gardens

If you have a low-lying area of the garden that is always wet and collects rainwater, then maybe it's best to just work with these somewhat soggy conditions, rather than trying to fight against them, through the creation of a bog garden filled with species that love the water. There is always the future risk that a bog garden will dry out in a drought, so make sure you incorporate plenty of compost or organic matter around the planting to enable it to retain moisture.

Green Roofs

Green roofs are increasingly common in our cities, and they have so many benefits. As we've previously mentioned, the one that interests us the most is the way they slow the flow of water off a roof and into the drainage system. They are the start of the sustainable drainage system (SUDS).

These natural living roofs are beneficial even if our rainfall becomes more intense because they reduce the volume of water going down drains. They also improve stormwater quality and provide building insulation, in addition to reducing the carbon footprint of the building and providing new habitats for wildlife.

In the garden, they can be built on sheds, firewood storage structures, and even chicken houses. They are relatively simple to install as long as you follow some guidelines. Note that a green roof installed over an occupied building or attached to an occupied building requires building permits. Check your local building codes before you get started.

First, you need to assess the existing roof. The slope is important; it should be no less than 2 degrees and no more than 10 degrees, any steeper and you run the risk of the roofing materials slipping, in which case you need to incorporate a grid structure to hold the substrate. Also, you need to work out whether the roof can take the weight! A typical green roof weighs between 12 and 31 pounds per square foot (60 and 150 kg/m2) when dry and will weigh far more when saturated with water, so make sure it's a sturdy roof with no holes, splits, or rotting wood.

A green roof is formed from several layers:

- A waterproof and root-proof membrane that is laid over the existing asphalt or shingle layer. This could be a butyl pond liner or a damp-proof membrane.
- A containment frame to retain the substrate, made from rot-resistant materials, such as treated wood or metal, with drainage outlets to allow water to drain away. Place a layer of pebbles on the membrane to ensure the drainage system does not get blocked.

A green roof on a firewood storage structure.

A green roof with a predominantly grass planting.

Substrate

This needs to be lightweight, low in nutrients, and moisture retentive, so don't use soil. Instead, turn to lightweight materials such as 75 percent crushed brick or expanded clay and 25 percent organic materials such as composted green waste. The depth can range from 3 to 8 inches (7 to 20 cm), with the deeper substrate providing greater drought resilience.

PLANTING

You can buy mats of sedum ready to roll out just like a new lawn, or you can place plug plants or use seed mixes designed for the dry and windy conditions of a rooftop. A diversity of species not only attracts more wildlife, but it is also more likely to become self-sustaining. When planting, bear in mind that species placed near to the top of the roof are going to experience the driest conditions. And be warned: some of the sedums can become a weed problem in the garden, popping up everywhere!

Rain and Cities

When it rains heavily in a city, the abundance of non-water-absorbent hard surfaces often results in a large volume of runoff water that the drainage system has to cope with. As stormwater from roofs and streets is directed underground and into local waterways, it carries with it debris and contaminants, which then end up in rivers and eventually the sea. This also means that the urban water table isn't being replenished with clean water but instead is often recharged with polluted material from storm drains and sewers.

It's no wonder that city planners and landscape architects are looking to incorporate sustainable drainage features, such as green roofs that retain water and slow down its rate of movement, as well as permeable pavements made using pavers, porous asphalt, and pervious concrete, which all offer an opportunity to put the brakes on the rate of water runoff.

Two decades ago, Toronto passed a bylaw that requires a certain percentage of each new roof construction over 21,500 square feet (2,000 m2) to have a green roof. In Canberra, Australia, water is harvested and recycled to help during periods of drought, and San Diego's Pure Water Program harvests rainwater with an impressive aim of producing one-third of the city's drinking water through recycling by 2035.

Rain gardens are appearing in cities around the world as part of soft landscaping schemes, helping to slow down the rate at which water runs off. In Tucson, Arizona, rain gardens have been installed along streets to slow, spread, and sink water that falls from the sky. Not only does this help during heavy rains, but the city also looks more attractive and the trees cast shade as a result. In Seattle, Washington, homeowners are offered incentives for the construction of rain gardens to aid the city's efforts in slowing down water.

An urban rain garden is functional, ornamental, and good for wildlife.

A tree pit provides sustainable drainage and slows water runoff, while providing the roots with water.

LEARNING FROM THE PAST:
Walls

The first recorded observations of the benefits of walls date back to 1561 when Conrad Gessner, a Swiss botanist, noted the effects of radiated heat from walls on fruit trees, enabling them to be grown in a more northerly location. During the 1600s, the Little Ice Age started to make its presence known, and around this time, walls and walled gardens became more important.

This isn't surprising when you consider that a thick wall could raise the temperature of the surrounding air by 19°F (10°C) at night, which meant that gardeners in northern England could grow a wide range of frost-sensitive fruits, such as peaches and apricots. By the mid-nineteenth century, Montreuil-sous-Bois, near Paris, had hundreds of miles of walls enclosing small spaces (clos) for growing peaches. Dutch gardeners opted for curved walls that were less thick and used fewer materials, but captured more heat and provided extra heat gain, while in eastern England, serpentine or zigzag walls were more usual. Often the walls were topped with overhanging copping stones for added protection, while projecting metal brackets allowed glass, net, or canvas to be hung vertically to protect plants.

The walled gardens of Montreuil-sous-Bois in the 1900s.

A serpentine wall (called a crinkle crankle wall in the United Kingdom) requires fewer bricks and traps more heat than a straight wall.

Living Walls

In the same way a green roof helps slow down the flow of water, so too can a wall covered with plants. Living walls can also improve the aesthetic appearance of the space, attract wildlife, keep a space cooler in summer, and provide extra insulation in winter.

There are small wall-planting systems you can buy for gardens, which are comprised of wall panels or modules filled with lightweight potting mix, connected to an in-built drip irrigation system. The types of plants that are suitable for these systems depend on the growing conditions. In sunny positions you could grow strawberries or bedding plants, while a shadier wall would be ideal for salad crops and herbs.

Garden walls

Walls are incredibly useful garden features. In effect, they create a localized, protected microclimate and assist in frost avoidance. South- and west-facing walls are particularly beneficial, taking up heat during the day and releasing it at night. This is useful in spring and autumn, when a boost of just a few degrees can be invaluable. In fact, it is estimated that the presence of a sunny wall is equivalent to moving 5 degrees latitude farther south, so the microclimate in a sunny, walled garden in Hampshire or Surrey can be likened to the conditions in Bordeaux, while a walled garden in Connecticut could be compared to conditions in Virginia.

South-facing walls are great for growing heat-loving crops like peppers and tomatoes, plus drought-tolerant Mediterranean plants because the wall provides heat, shelter, and protection from the extremes (especially wind and frost). Do remember it can get very hot too, so water and mulch well. If you are worried about frost-sensitive plants in pots, move them beside a wall on a cold night and cover them with a row cover. You can also construct a temporary lean-to with shelves for trays of sown seeds and young plants and drape a row cover or plastic over the front, which should be enough to protect these tender plants from frost and heavy rain.

A living wall slows down the flow of water.

RIGHT: **Sally grows fruit trees along a southwest-facing wall while the nearby vegetable beds also benefit from the microclimate created by the high wall.**

WORKING WITH WILDLIFE

What should a garden look like? Although more naturalistically minded planting has increasingly been embraced in recent years, there is still a very much ingrained perception that our outside spaces should be neat and nature firmly kept in check. There may be wild edges, but the ideal garden from this viewpoint probably has immaculate weed-free, well-trimmed short lawns, trees and shrubs uniformly pruned back into place, exceptionally neat, symmetrical flower beds, vegetable patches full of blocks or rows of produce . . . and all of it managed with meticulous detail for most of the year.

During the summer months, much time is spent primping and pruning to keep everything in its place, and at the end of the season, as the daylight hours start to shorten, preparation for winter begins in earnest. Old annuals are pulled out, the soil is most likely dug over to expose it to the elements, messy leaf litter is cleared away, and all semblance of the natural cycle of decay removed entirely from view.

LEFT: We need to encourage essential pollinators such as bumblebees.

Butterflies are no longer a common sight in gardens.

A wildlife pond can attract dragonflies and damselflies.

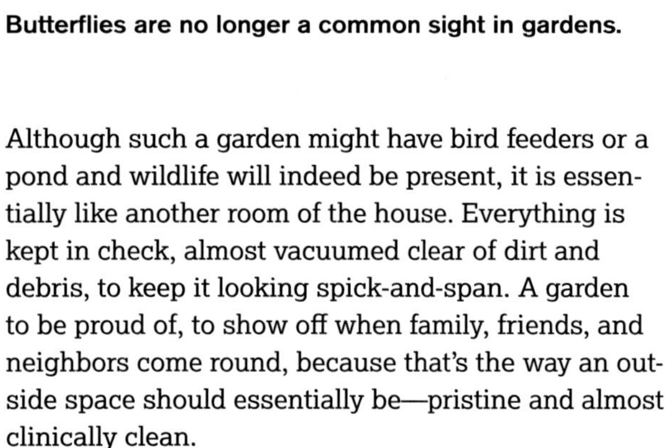

Although such a garden might have bird feeders or a pond and wildlife will indeed be present, it is essentially like another room of the house. Everything is kept in check, almost vacuumed clear of dirt and debris, to keep it looking spick-and-span. A garden to be proud of, to show off when family, friends, and neighbors come round, because that's the way an outside space should essentially be—pristine and almost clinically clean.

Not only is it an awful lot of backbreaking work, but also the result of keeping an outside space in such a spotless condition means the plants are a lot more vulnerable to the elements and in need of year-round attention. As you know by now from the chapter on soil, digging your garden over and leaving it weed-free over winter is damaging, making your precious soil much more susceptible to nutrient loss. Likewise, in the summer months, a neatly ordered and spaced flower bed is going to dry out a lot more quickly than one that is wilder in aesthetic, with much more ground cover to keep precious moisture in.

Alongside the many, many time-saving aspects of letting your garden grow a little wilder (removing the need to replenish the lost nutrients from the eroded soil with fertilizers and improvers for one, and reducing the back-and-forth watering requirements during the summer for another), the benefits to wildlife are immense. In turn, the advantages to your garden and to you, as the custodian of your plot, are tenfold when it comes to helping to build a natural biodiversity and resilience from within.

The truth of the matter is that building natural biodiversity in your outside space will be a key tactic against the more volatile weather of the climate-changed future. A much more naturalistic, low-maintenance space is less work for you as the gardener and is better able to withstand the vagaries of our future weather.

Future Pests and Diseases: The Threat

With milder winters, more pests are able to survive and prosper. For example, for every 1.8°F (1°C) increase in average temperatures, aphids will become active two weeks earlier than usual. Plus, according to scientists in the journal *Science*, as temperatures rise, so do the metabolic and reproductive rates of insects. On the ground, this means we're facing the prospect of hungrier insects in greater numbers, and there's also the risk of new species moving in and prospering as a result of the changing climate. We'll also see more fungal disease, such as powdery mildew and black spot, and overall there will be an increase in the number of threats faced by the plants we grow. The following chapter explains this in more detail.

Letting nature in to lend a helping hand also helps deal with the increased threat from new and unusual pests that loom on the horizon. A garden that is in balance with nature, and therefore has a wide diversity of plants and animals within it, is a much harder place for individual species to dominate or create problems. Such a biodiverse organic space provides a natural defense that money simply cannot buy, in addition to making your garden a truly mesmerizing and enticing place in which to be. Imagine being able to spot frogs and newts in the vegetable patch as you garden, seeing ground beetles and other enthusiastic slug-eating predators in abundance, listening to the sound of bees working the pollen on your plants, and watching almost spellbound as a dragonfly comes into view on a warm summer's day.

All creatures have a role to play as part of the natural balance. So, by letting go of this image of the perfect, manicured plot and actively encouraging wildlife in, our workload will be lessened, our gardens afforded greater protection, and our lives enriched beyond measure.

Changing Seasons

One element of climate change that can't have escaped anyone's notice is the early arrival of spring. Here in the United Kingdom, we've seen daffodils at Christmas and snowdrops in November. These "early arrivals" vary, depending on where you are in the world. For every 10 degrees north from the equator, spring now arrives around four days earlier than it did 109 years ago, and this change is happening at a far faster rate than previously thought. In Los Angeles, spring arrives a day earlier than it did in 2008, but in Seattle it's four days earlier. In 2017, spring in Washington, D.C., was twenty-two days early. And most worrying of all, one of the most extreme changes is observed in the high Arctic, where spring is arriving as much as sixteen days earlier. And it's not just that the timings are earlier but that there is far more variability, so one year spring could be incredibly early and the next it could be surprisingly late.

These seasonal changes disrupt natural cycles—the migration of birds and amphibians, the pollination of flowers, the availability of food plants for insects, and many other ecological relationships. There are other side effects too. Ticks and mosquitoes remain active for longer, the hay fever period goes on and on, and plants are at greater risk from frost in spring or from a summer drought.

Snowdrops are appearing ever earlier.

Changing weather patterns affect insects in a variety of ways, including making it harder for them to find food. Stress also makes them more susceptible to disease, as we see with the spread of varroa mites in honeybee populations. A milder winter or earlier spring can favor pests that come from other parts of the world (see chapter 7). The spotted-wing drosophila, for example, is now spreading quickly across the United Kingdom. One way to help is to grow a wide range of nectar plants with a long season so there is always something for pollinators to find in times of need.

The Beneficial Role of Paper Wasps

Yes, they can be the bane of a late-summer barbecue, zooming in on any sweet-smelling drinks or meaty foods almost as soon as you've laid the picnic table. They also pack a powerful sting if you're unlucky enough to be the recipient of their angry attention. Yet, at that time of year, when they have a well-known reputation for aggressiveness, paper wasps are in fact coming to the end of their life cycle and slowly but surely dying. I think that would make the best of us more than a little grouchy.

Also, it's worth bearing in mind the vital role wasps play in the ecology of our gardens. They are incredibly useful predators, hoovering up aphid infestations on crops and helping to keep many types of pest caterpillars in check on our behalf. Wasps are part of the natural order of predators, keeping other insect numbers in balance. They also play an important role in the pollination of our plants.

Encouraging Creatures in Your Garden

One thing is certain: if you want to improve the climate resilience of your garden, you need more biodiversity, both in the soil (see chapter 4) and above the ground. Ultimately, if you allow even one small area of your garden to become a little unkempt, letting wildflowers move in (even in pots), the grass lengthen, and a few weedy species prosper and grow, that alone will encourage a greater range of wildlife in your outside space. Add a small pond or water source, a bird feeder or two, and some stones or a wood pile—and grow a wider range of plants organically—and you will see your outside space truly come alive.

Attract Pollinators

When it comes to pollinators, we're not just talking about bees. Many other animals participate in the process of pollinating your plants. From wasps and hoverflies to butterflies, moths, flies, and even bats, there is a range of creatures that carry out this vitally important role.

The greater the mixture of flowering planting you have, the better your garden will be from a pollinator perspective. During the summer months, most gardens will have an ample supply of food, but having energy-giving nectar on tap early in the season can make all the difference between survival and suffering for bees and other insects that are emerging from hibernation. The same is true when it comes to building up supplies late in the season to enable insects to overwinter successfully. Late-season nectar provides them with an enhanced ability to deal with adverse conditions and fend off disease, whether they are a species that live in a colony or singularly.

The flowerheads of angelica attract insects.

FLOWERS

To ensure as wide a diversity of pollinator-friendly plantings as possible, work with flowering plants native to your area that are in bloom at different times of the year. Look around your community for ideas and inspiration. From noninvasive wildflowers early in the season on waste ground to hardy perennials and shrubs in neighbors' gardens, see what fares well locally and weave it into your outside space with gusto.

Allowing brassica to flower early in its second year will attract bees galore.

HERBS

The likes of lavender, sage, rosemary, thyme, marjoram, chive, and mint will draw bees from far and wide to your garden during the summer months. Beekeepers often recommend these herbs if you're going to grow anything for bees in pots or hanging baskets.

FRUIT AND VEGETABLES

While pollinators are necessary for much of the vegetable garden, so the vegetable garden is useful for pollinators. From apple trees and stone fruits to beans and summer squash, the flowers of our crops are much-favored sources of nectar and pollen. Allowing some of your brassicas and root crops, parsnips in particular, to overwinter and set flower in their second year is a good way of attracting pollinators and predatory beneficial insects to your plot early in the season. The pretty yellow flowers of parsnips will buzz with pollinator activity. Leek flower heads are also much favored later in the year. Generally, allowing some of your produce to flower in this way provides a useful source of food for insects, as well as enabling you to save seed.

Provide Habitat

As well as providing food plants, it's also important to help pollinators overwinter successfully on your plot. While honeybees will come from a hive up to 3 miles (5 km) or so away, bumblebees and most solitary bee species tend to nest much closer to their food sources. They like to bury themselves in something, such as a dead or decaying branch on a tree, a wood pile, a pile of leaf litter, or in your compost heap. Providing suitable areas for them to find some shelter against the winter elements will enable them to emerge directly where you need them most come early spring.

AMPHIBIANS

These delightful creatures are another boon to the garden, providing an extremely important service in myriad ways. For a start, they actively hunt and eat slugs and snails in your garden. As carnivores, they eat whatever they can get their hands (or rather tongues) on, including caterpillars, mosquitoes, flies, beetles, spiders, wood lice, ants, and, in the case of large toads, sometimes even small mice!

Frogs are vegetable garden heroes.

A pond of some form is the most obvious way to attract these creatures. In fact, any water source can be effective when it comes to tempting frogs and toads into your garden (see sidebar). Equally as important, however, is ground cover, both around the pond itself and along corridors within the garden. These areas provide space for frogs and toads to seek shelter from the midday sun and offer some protection from larger predators, such as cats. Also, it's important not to use any chemical products or pesticides in the garden because amphibians are very sensitive to such toxins.

HEDGEHOGS

Found across Europe, the European hedgehog (*Erinaceus europaeus*) loves wild corners of the garden where they can find somewhere to rest, nest, and hibernate. Log piles, compost heaps, and piles of leaves all provide a warm, overwintering spot. These nocturnal visitors truly are the gardener's friend, eating slugs, caterpillars, beetles, and other invertebrates.

The hedgehog is a true friend of the gardener

Be Careful When You Mow

During the summer months, amphibians tend to take shelter in long grassy areas, so it's important to be vigilant when mowing the lawn. The same goes for moths, as they often find this an attractive habitat during the day. Leaving some areas of grass to grow long affords these beneficial creatures some vital protection.

A compost pile provides an attractive home for amphibians to overwinter (which is why you don't want to turn a heap during the winter months). Another alternative is a log pile or an old terra-cotta plant pot raised slightly above the ground to provide vital shelter for toads. A rockery with lots of nooks and crannies to hide in is another desirable option for frogs. Newts and salamanders will bury themselves in the ground or find a nice crack in a wall to hide in over winter, so it's important to be especially careful come spring when working the soil. I often come across these little critters buried in the soil in the polytunnels, so I work the soil gently with my hands to ensure they come to no harm. If I were to use a spade to turn the soil, I would very likely to kill or injure them. This is yet another good reason for not turning the soil.

Despite its huge popularity, the hedgehog is in decline across its range, especially in the United Kingdom. This is mostly a result of habitat loss due to changes in farming practices. As of this writing, their numbers are down by as much as 50 percent since 2002 and gardens have become more important for their survival.

BIRDS

It's impossible to imagine a garden without a varied assortment of feathered friends. As well as making our outdoor spaces so much more enjoyable, they act as useful allies in the garden, eating troublesome pests, such as slugs, snails, aphids, and caterpillars, among many others. Feeders and a water source will, of course, attract birds, but it's also important to think about plants that will provide more food and places to shelter. Leaving seed heads in place over winter benefits birds as much as the soil, while hedging plants such as hawthorn and holly also provide a source of berry-powered winter fuel as well as a nighttime roost in winter.

Building a Bug Hotel

One way to ensure your garden contains a variety of mini beast habitats is to build a bug hotel from old pallets, bits of wood, old bamboo canes, bricks, clay pipes and pots, and anything else you can find. The nooks and crannies of such a habitat are the perfect hiding places for insects, spiders, centipedes, and even amphibians. The more varied the habitats you create, the more biodiversity you will attract. Use old pallets to create layers (floors) and then fill the gaps with your recycled materials; the more varied the materials, the better. You can finish off with a roof or even create a green roof for your bug hotel (see page 93).

Look out for ladybug larvae.

Encourage Predators

Many animals in the garden have the valuable role of predator, helping to keep certain pests in check. Look no further than ladybugs (ladybirds), hoverflies, and lacewings. This fantastic trio produce larvae that will eat their way through a colony of aphids like there is no tomorrow. They all play a vital role in keeping pest numbers in check.

Ladybugs and lacewings also eat many other pests, while adult hoverflies play an important role in plant pollination. Plants to entice these predators include members of the daisy family, dill, and fennel, among others. In fact, one organic vegetable producer we know always picks a large bunch of fennel and takes it into one of his polytunnels if he's had an aphid outbreak to allow the ladybug and lacewing larvae living on it to work their magic.

Did you know ladybug larvae don't look anything like you'd imagine? Nor do they look anything like the adult bugs. They are so incredibly useful in the garden, it's worth knowing what they are to ensure you allow these valuable natural pest control allies to prosper.

Parasitic wasps are also drawn to fennel. These valuable wasps lay their eggs in the body of caterpillars (yuk!) and certain other insects. When they hatch, the wasp larvae literally eat their host insect from the inside. Then they pupate and emerge as new wasps.

Other essential predators in the garden include spiders and ground beetles. You may jump back when you lift a pot and a black beetle scuttles away, but this fast-moving insect is a voracious predator, chasing down prey on its long legs.

Build Your Own Wildlife Pond

Wherever you live, whether it's in the bustling heart of a city or in the quiet countryside, if you create a water source outside, you're almost guaranteed to draw wildlife in. Even the smallest water feature can attract a wide range of beneficial creatures to help provide a source of interest and wonder, as well as making sure your garden truly comes alive.

There's nothing like listening to birdsong as you sit or work outside, and such feathered friends will soon be drawn to any opportunity for a little drink or splash around in water that you make available to them. In return for refreshments, they'll most likely reward you by dealing with any slugs, snails, or aphids they happen to find causing trouble on your nearby plants. Much like other water-loving creatures, such as frogs, toads, newts, and dragonflies, birds provide a very useful and equally mesmerizing addition to any outside space.

Building the Pond

The first step is to find a suitable liner, preformed plastic pond, or container. If you opt for a pond liner, there is a huge range of butyl rubber liners available nowadays in a wide range of sizes, and they are very easy to work with. Choose an area with sunlight but also some daytime shade. To make it tempting to as much wildlife as possible, some privacy is preferable, so siting it in a corner or at the back of your garden is ideal.

It's perfectly feasible to have a raised water feature, too. For a raised patio pond, positioning doesn't matter so much; any water source is beneficial, so just choose where it would fit in best—perhaps nestled somewhere on your balcony or front yard, for example. However, amphibians will struggle to get in and out of a raised pond, so it is always preferable to have a pond that is sunk into the ground. To do this, you will have to dig a big hole in your chosen spot. You could meticulously measure the area first, or just figure it out as you go along; you'll get there either way. Ideally, you want the pond's liner sitting either just below or level to the ground around it.

It's important to provide access for wildlife, and you can do this using some of the natural materials you have at hand. The aim is to make it easy for frogs, toads, and newts, in particular, to get in and out. Pebbles, placed in and around the pond, do the trick nicely and have the added benefit of providing places under which to hide. Planting around the area also provides welcome ground cover and will work to make the space even more amphibian friendly. Plus, if you carefully position stones, rocks, or wood in the pond itself to create shallow edges, this makes it easier for the birds to bathe, too.

Create a pond and watch the wildlife move in.

It's best to fill your pond with rainwater because tap water is chlorinated, so the contents of a rain barrel would be ideal. Otherwise, you can just leave it to fill naturally over time.

Keep the pond in good health (and the water clear) with the addition of aquatic plants. These help aerate the pond and are a good source of food and shelter for pond life. If the water level dips over time, simply keep it topped up with collected rainwater.

New Visitors

As the water settles, birds are likely to move in first, enjoying the opportunity for a drink and maybe even a bath. Over time, as the pond matures, you're likely to see all manner of wildlife. Frogs and salamanders take up residence in even the smallest of ponds, especially if they have shelter nearby.

("Build Your Own Wildlife Pond" was first published in *The Primrose Water Feature Book* by Kim Stoddart and published by Primrose.)

PESTS, DISEASES, AND ALIENS

Climate change brings other threats to our gardens. It's not just extremes of climate that we have to cope with, it's also the arrival of new pests and diseases, plus competition from invasive plants. As the world warms up, pest and disease organisms can expand their range and travel further afield than they would do otherwise. With these widening territories, pests can become highly competitive and invasive once they set out establishing themselves in a new area.

In addition, warming temperatures, and especially milder winters, enable some pests to overwinter, emerge, and breed more rapidly than they have previously, resulting in greater numbers overall. For example, in the northeastern United States, Mexican bean beetles, Colorado potato beetles, and others might have produced just one generation of offspring in the past. Now, two generations per season is sometimes the case. With potentially greater pest numbers, pests arriving earlier in the growing season, and new emerging threats, this can make it more challenging for the gardener to keep pest numbers and their impact in check over a longer period of time.

Over the last few decades, globalization has led to more international trade in plant materials, including wood, foods, ornamental plants, and more. This has resulted in insects being unwittingly introduced to areas outside of their normal range. In the past, these insects may not have been able to get a foothold, but our changing climate means that sometimes they now can. Diseases too can be spread around the world via infected timber or ornamental plants.

LEFT: Iconic trees in our cities are under threat.

Invasive insects are a threat to natural ecosystems and agriculture, with concerns around food security and potential economic losses through crop failures. It's a worry to gardeners also. Certainly Henrik Sjöman is anxious. He is a researcher from Sweden and the scientific curator for Gothenburg Botanical Garden. He says that more than half of the trees growing in northern European cities are of just two or three species, so if they prove vulnerable to attack by pests and diseases, the urban landscape will be changed, and not for the good. He quotes one example: Helsinki in Finland, where almost half the trees in urban streets and parks are linden trees (*Tilia × europaea*). And other cities around the world are not much better, with just a few species making up the bulk of the urban trees.

How Species Are on the Move

It's common to get migrant species arriving at a certain time of year—butterflies in summer or waterfowl in autumn, for example. In normal times, these creatures would either make the return journey to warmer overwintering locations or die when conditions are not suitable for them to survive. But as the climate warms up and conditions change, they may be able to survive all year-round, which can lead to resident populations rather than seasonal visitors. This might be fine if it is a useful insect such as a butterfly, but what if it is a pest species whose numbers would ordinarily be kept in check by cold winter temperatures and seasonal dieback?

The key drivers are temperature and rainfall. Insect pests like warmth so they can be active. This means that increases in temperatures and rainfall tend to favor their growth and distribution potential. However, on the plus side, very high temperatures will slow them down and heavy rain may wash away their eggs or larvae. Researchers are finding that pests are moving away from the tropics, where extreme weather is making itself felt, and toward more temperate zones. They are moving relatively quickly, too, at an average of 2 miles (3 km) a year toward the north and south poles. For some organisms, it can even be much faster because they are also carried by winds.

The changing climate is favoring fungi too. The higher temperatures lead to more spore formation and therefore a greater chance of infection from invasive pathogens such as blight. Fungi evolve quickly, so new strains may appear. Carbon dioxide has a role to play too. As carbon dioxide levels in the air rise, the plant performs more photosynthesis, and often this leads to more leafy growth, disproportionate to plant size, which can make the plant more vulnerable to pests and disease.

Impact on the Ground

As much as 40 percent of the food crops produced around the world each year are already lost to plant pests and diseases, and experts are expecting this figure to increase as a result of climate change. Research by Curtis Deutsch at the University of Washington looked at how climate change could affect thirty-eight of the world's most studied insect crop pests. Rising temperatures increase the rate at which pests digest food, so they can munch their way through a crop more quickly. Warming climes can also enable insects to be active for longer and move around more easily to find crops and breed.

If global temperatures were to rise by 3.6°F (2°C), pest-related crop losses are estimated to increase

The Marmalade Hoverfly

The marmalade hoverfly (*Episyrphus balteatus*) is an important pollinator and predator in gardens across Europe. It is a migrant species that overwinters in North Africa and then flies north in spring. More than 4 billion of these hoverflies are estimated to reach the United Kingdom every year, where they spend the summer pollinating flowers while their larvae munch aphids. In autumn they return south. As a migrant species, the marmalade hoverfly is often better able to find suitable habitats than some of the endemic hoverfly species. It may also benefit from global warming, as milder winters will allow it to survive northern European winters without the need to migrate.

by 46 percent in wheat, 19 percent in rice, and 31 percent in maize compared with current levels at the time of this writing. Rice is least affected because it is grown in tropical areas where temperatures are already high. Temperature increases will provide the greatest benefit to the insect pests living in Arctic, boreal, and temperate zones. Amazingly, just a single unusually warm day in winter could be enough to help the establishment of some invasive species.

The Winners and Losers

Every region of the globe is seeing impacts due to the changing climate, and there will be both positive gains and negative outcomes for species across the board. Some butterflies, for example, are benefiting from the changes and are becoming more commonplace and expanding their distribution, while others are not faring so well.

Species that have more general requirements are at an advantage. A weed, for example, has a greater chance of expanding its range because it is adaptable and able to colonize bare ground if conditions suit. It has a short life cycle and produces lots of seed to maximize its chances of spreading, whereas a plant with a specific soil requirement—an ericaceous plant, for example—wouldn't be able to spread so easily.

It is the same among the insects. A butterfly that feeds on a range of flowers and lays its eggs on numerous hosts will be better placed to take advantage of new territories than a specialist that has an intimate relationship with just one species of plant. In the case of the specialist, if climate change affects its habitat, chances are it will decline or even disappear altogether. Given their characteristics, it is not surprising that insect pests tend to be generalists and opportunistic. Although some pests will only attack a specific host plant, most are able seek out a whole range of plants.

It only takes a couple of mild winters for an insect pest, previously restricted by the cold, to have the chance of overwintering successfully in an area beyond its normal range. Then it has the chance of an early start to the year because as soon as the temperatures in spring warm up it can become active and start breeding, whereas before it had to fly north.

The Case of the Longhorn Beetle

Longhorn beetles are well-known forestry pests. Two in particular are causing concern: the Asian long-horned beetle (*Anoplophora glabripennis*) and the citrus long-horned beetle (*Anoplophora chinensis*). Both are highly adaptable species from China. In their native forests, they are kept under control by natural predators, but as they spread across the world, helped by climate change and international trade in timber and plants, they lack that predation and pose a significant threat. The Asian long-horned beetle appeared in North America in the 1990s and since then has caused widespread damage to forest trees. Worldwide, it's known to attack more than 130 different species of tree. Individual citrus long-horned beetles have been found in Europe and the United States, but so far there has been no outbreak. But this beetle is known to attack at least 188 different species of tree, and it is spreading west from China. According to Henrik Sjöman, should they get established in northern Europe, citrus long-horned beetles have the potential to wipe out most of the urban trees in ten major Nordic cities if a worst-case scenario proves true. The lack of diversity in urban tree species makes the threat worse. And garden trees would not escape the damage either.

The citrus long-horned beetle.

Often a pest will not have any natural predators in its new range, so for the first years at least it can survive without being controlled by predators.

When a tree species declines through climate change or disease, it is not just that one species that suffers, but the many species of animal that rely on that tree as a habitat, food source, and so on. An oak tree may be host to hundreds of other species—insects, birds, mammals, mosses, fungi, lichens, and more. If the tree goes into decline, there is a residual effect on these other species too.

New Pests and Diseases

Plant health specialists look at pests and diseases in other countries and work out which ones are the most likely to cause problems in the future. Here are a few of the pests and diseases that are causing concern.

Pest and Disease	Characteristics and How to Deal with the Threat
Allium leaf miner (*Phytomyza gymnostoma*)	This small fly affects alliums of all species. It was first reported in Europe in 1976 and in Pennsylvania in 2015 and Massachusetts in 2019. There are two generations, spring and late summer to autumn. Allium leaf miners overwinter as pupae and emerge as adults in spring. They lay eggs in leaves and then the larvae tunnel into the stem. The larvae pupate in the stem or surrounding soil and emerge later in the year to attack autumn crops. They are difficult to treat with chemicals, so the best controls are insect nets over the crop when the flies are active.
Boxwood blight (*Cylindrocladium buxicola* and *Pseudonectria buxi*)	This fungal disease has been ravaging boxwood plants across Europe since the 1990s, helped by milder and wetter winters. The blight causes leaves to turn brown and fall off, leaving brown patches in boxwood hedging. The disease spreads quicky and as yet there is no treatment, other than removal of infected branches. Measures gardeners can take to keep the disease out of their spaces include buying less susceptible cultivars, quarantining new plants, not growing boxwood in damp and shady areas, not overcrowding the plants, avoiding overhead watering, and ensuring general good garden hygiene to avoid spreading the fungus.
Mountain pine beetle (*Dendroctonus ponderosae*) and southern pine beetle (*D. frontalis*)	Weather patterns such as drought, warmer winters, and storms have allowed these beetles to expand their range, posing considerable threat to pine forests. The mountain pine beetle has destroyed large areas of pine forest in the U.S. Pacific Northwest, while the highly destructive southern pine beetle is moving north as temperatures rise and is likely to spread throughout the northeastern United States and into southeastern Canada by 2050.
Southern blight (*Athelia rolfsii*)	Southern blight is a fungal disease that affects a wide range of vegetables in tropical and subtropical regions of the world, including Florida, where it was first discovered. It is spread easily on soil and bulbs. As temperate regions get warmer, conditions will favor the spread of this fungus, which is already common in parts of Europe.
Xylella (*Xylella fastidiosa*)	This bacterial disease is carried by sap feeders, such as leafhoppers and spittlebugs. The vectors' feeding habits allow the bacteria to gain entry to the plant, where they block xylem vessels and reduce water flow, so the plant suffers from leaf scorch, wilt, and dieback. Once found across the Americas, including the southern United States, this disease has now spread north. The disease appeared in southern Italy in 2013 and has since devastated olive groves in Italy, France, and Spain. Unfortunately, there are a huge range of host plants, including oleander, grapes, lavender, coffee, citrus, and many species of hardwood tree. The disease benefits from climate change because the insect vectors are more active in warmer temperatures, while the higher temperatures stress the host plants, weakening them. This makes them more vulnerable to attack. Countries have various controls to stop the spread, which involve import restrictions, quarantine measures, and, in the case of an outbreak, the destruction of all host plants.

Alien Species

Invasive alien species are "animals, plants, or organisms that are introduced into places outside their natural range and negatively impact the native biodiversity or ecosystem services," as defined by the International Union for the Conservation of Nature (IUCN). They are one of the biggest causes of biodiversity loss.

One of the best-known examples is the water hyacinth (*Eichhornia crassipes*), a South American plant that forms a dense, impenetrable floating layer over a water body, blocking out light and making leisure activities, such as boating, impossible. It was introduced to the United States in 1884 as an ornamental aquatic plant but escaped from ponds and now colonizes waterways throughout the southeastern United States, where it causes huge management issues. It's causing problems in other places too, including Lake Victoria in East Africa.

Other examples include Himalayan balsam (*Impatiens glandulifera*), which was first introduced to the United Kingdom in the 1850s and now has a widespread distribution, and rhododendron. More problematic is Japanese knotweed (*Fallopia japonica*), which arrived in Europe in the 1800s as an ornamental plant and has now spread across the continent. It's a major concern because it spreads quickly to form dense stands several feet high, which are difficult to remove, and the rhizomes can grow through asphalt and concrete.

Invasive species by their very nature are opportunistic and generalist, so they can cope with a wide range of environmental conditions, grow in many different soils, have good seed dispersal, and boast a fast growth rate. This means they are in an ideal position to take advantage of milder winters and warmer summers that climate change will bring to temperate regions.

And we must not forget that many ornamentals grown in our gardens are considered to be invasive in other parts of the world. In the United States, the top five invasive species according to James Gagliardi, horticulturalist at the Smithsonian Gardens, are purple loosestrife (*Lythrum salicaria*), Japanese honeysuckle (*Lonicera japonica*), Japanese barberry (*Berberis thunbergii*), Norway maple (*Acer platanoides*), and English ivy (*Hedera helix*), species that are common in European gardens. More than half of Europe's currently naturalized and invasive alien species are garden escapes, so ornamentals are a concern when we think about the future.

RIGHT: **Water hyacinth is a floating species and grows quickly, often blocking waterways.**

English versus Spanish Bluebells

In 2008, English bluebells (*Hyacinthoides non-scripta*) (shown at right) were recorded as flowering in February rather than April and May. Studies suggest that this species is sensitive to spring temperatures, and it will come into flower or leaf between three and eight days earlier for every 1.8°F (1°C) increase in temperature. It was named as one of the four UK species most likely to struggle with climate change, the others being garlic mustard (*Alliaria petiolata*, a species that is a noxious invasive weed in the United States), larch (*Larix decidua*), and sycamore (*Acer pseudoplatanus*). Those best placed to cope with the changes are ash (*Fraxinus excelsior*), beech (*Fagus sylvatica*), silver birch (*Betula pendula*), and wood anemone (*Anemone nemorosa*).

The Spanish bluebell (*Hyacinthoides hispanica*) was introduced to gardens in the late seventeenth century. It's a vigorous garden plant that produces scentless, pale blue flowers with paler stripes. It has now become naturalized in hedgerows, woodlands, and parks and is spreading quickly. Climate change suits it, so its spread is likely to continue as temperatures rise.

Many landscape architects are looking to the flora of other countries to find species suitable to future conditions. For example, they may seek out plants from South Africa or Australia to grow in temperate regions. These plants may not be a problem in their native environment, but we don't know how competitive they may be in a new one. Scientists will look at a species' climatic niche to determine whether there is a risk of naturalization. In the past, a Mediterranean species that was not cold hardy would not be considered a risk, but climate change may bring mild winters and therefore the risk may change. So now the risk assessment has to include other features. For example, a potentially risky grassland species may be a perennial, as its more likely to germinate in undisturbed grassland habitats than the seeds of annual species. This is because annual species are less competitive when it comes to getting a foothold in an existing sward, while a species that produces a heavier seed may see greater survival when colonizing undisturbed plots when compared to a species with lighter seeds.

How Will We Cope with These Threats?

The answer lies in diversity of plantings and working with wildlife for biodiversity overall. We need to plant as many different species as possible so that our urban trees and garden plants are more resilient to the pests and diseases that will come with climate change. This highlights the need for landscape architects and gardeners to grow a much greater diversity of plants and trial new species that are not in the nursery lists.

As you read in the previous chapter, nurturing wildlife and encouraging predators also helps redress the natural balance. There are many ways to expand and diversify your plantings to help build resilience. In the next chapter, we explain how to build a low-maintenance and productive vegetable patch that has a better chance of thriving whatever weather, pest, and disease threats lie ahead.

THE VEGETABLE GARDEN

We live in very uncertain times, and food security in the future is going to be impacted dramatically by the changing climate and rising cost of living. As consumers reliant on buying in all of their food for the table, this leaves us open to the volatility of what is a global commodity market. A flood or drought in one part of the world, decimating harvests of certain produce, can have a dramatic ongoing impact on supply and prices elsewhere. The same goes for demand. For example, a prolonged heat wave, when demand for summer salad will be high, can conversely see home vegetable growers struggling to keep up with demand, as lettuce simply stops growing in temperatures above 86°F (30°C). If the only available supply sources are overseas, as imports increase, so will prices.

Our changing climate is most likely to make food shortages, supply chain delays, and soaring costs a distinct part of our future. When you consider this—alongside the undeniable fact that out-of-season produce is often a sorry excuse for freshly picked sweetness—the case for growing your own is rock solid. There's no plastic packing, no shipping or haulage, and no unknown chemicals to worry about, just a great feeling as you consume the vegetables you have lovingly grown. Growing your own food is worthwhile on many levels, and there is yet another benefit. If you improve the resilience of your garden to extreme weather events and the threat of new pests and diseases, you will be rewarded with the ability to grow a greater variety of plants and vegetables all year-round.

LEFT: A biodiverse planting on Kim's vegetable patch.

Change Is in the Air

Many of our current grow-your-own practices are high maintenance and vulnerable to the elements, based as they are around a set of (let's face it) rather exacting dos and don'ts pertaining to the cultivation of homegrown fruits and vegetables.

A lot of this comes down to an ingrained desire for uniformity—be it straight lines or blocks and blocks of produce that you see when you travel to any community garden around the world. We're programmed to keep our veggie plots looking very neat and tidy, to keep nature very much in its place and under control. Yet, as you read in the previous chapter on wildlife, actually letting nature in to lend a helping hand makes for a much hardier, low-maintenance, productive, and, arguably, more enjoyable place in which to grow. It encourages the natural biodiversity of nature. It means the difference between managing a plot that is truly alive, in which plants, the soil, and the creatures all work in balance together, and a sterile, high-maintenance plot that has exacting demands because it's kept so clinically clean.

Understandably, jumping wholeheartedly into a different way of gardening and growing food may be too much for some people to take on all at once. Yet shoring up your defenses, even little by little, will enable you to weather the extremes that we face in the future.

Creating a more resilient garden means, in part, becoming a more resilient gardener. The two go hand in hand, and that will take time and confidence. It's really not rocket science; it's just an ability to look outside of the traditional gardening rule books and instead tune in to your self-instinct, as well as the natural potential of your outside space.

When to Grow

With increasingly changeable weather, we can no longer predict when to grow plants based on a "standard timetable." Although many gardeners undoubtedly take comfort from certain crops being sown, regular as clockwork, at a certain time of year, such a disconnected approach is becoming increasingly unreliable. Nowadays, it's vital to look at what is happening around you, at what the weather is doing, and to adopt a more flexible approach so that "good old common sense" dictates when will be the best time to sow your precious seeds.

It's worth changing these patterns of planting anyway, as the ability to adapt, and to swap and change where needed, will be key in the near future. Ultimately, we're looking at a future where both early and late sowings are likely to be possible as we move toward a two-season pattern of growth, with spring to early summer and late summer to autumn. It's a pattern that's more typical of the Mediterranean.

LEARNING FROM THE PAST:
An Organic History

The gardens of old were rarely decorative. Instead, they served a purely functional use, especially for those in so-called lower status positions, as the garden was used to provide food for the table. Within this system, some weeds were actively encouraged because of their many uses in the home. In addition to protecting the ground against nutrient leaching over winter, many weeds that we would nowadays remove, such as chickweed (*Stellaria media*), lamb's quarters (a.k.a. fat hen), and ox tongue (*Picris ecchioides*) would have been encouraged to self-seed and grow in between crops. This would have served the purpose of attracting beneficial insects to aid natural pest control, as well as providing a welcome food source, especially during the "hungry gap" in spring. The ground was always kept covered with plants, helping to keep moisture in during the warmer months and avoiding erosion in winter.

Herbs also played a very important role, with a wide range being grown for myriad practical uses, from flavoring food in the kitchen and dyeing textiles, to filling beds and bedding (wormwood et al. [*Artemisia absinthium*]), keeping out bedbugs, and even warding off evil spirits (rosemary [*Salvia Rosmarinus*]). These plants were also used for a range of medicinal purposes (such as comfrey to help mend broken bones). Seed-saving was another important part of the gardening calendar, with plants harvested at the end of the season to provide the source of planting for future years.

Flowers were only grown in the highest status gardens at first, with plant finders often sponsored by wealthy gardeners to seek and find interesting species to be added to the plantings on their estates. It was only over time that these species would have filtered down to more everyday use in less prestigious gardens.

Although most medieval gardens followed a three-year rotation system of peas and beans, followed by grain and then a fallow period, a number of crop seeds were often sown together to reduce risk of failure. This would have resulted in a mixed planting that, combined with the ample wildflowers and weeds grown alongside, created a healthy degree of biodiversity and natural resilience.

Building Resilience in the Gardener: A Change in Mind-Set

This book gives lots of advice, information, and ideas to protect and prepare your garden (and the plants therein) for the climate change extremes of weather ahead. We have already established that it's no longer gardening as usual, and while this undoubtedly presents challenges, it also offers great opportunities. There's no turning back the clock, and the development of personal resourcefulness inside better enables us to deal with the hostilities on the outside.

What might help in this process of adapting to new ways of gardening is the knowledge that much of what we consider traditional advice is based on the comparatively recent garden care practices of large, European country estate houses. In those days, activities and the calendar were designed for a team of gardeners working at the behest of the high-ranking owner. Everything was designed to keep the gardens primped and polished, and the kitchen full of produce, for the estate owner's pleasure.

Also, there was the not-so-insignificant matter of keeping the staff busy all year-round so there was no slacking off. As such, work generation was certainly part of the equation at quieter times of year, with the creation of time-consuming practices that are still dutifully carried out to this day.

Prior to this very controlled approach, in which outside spaces were kept meticulously tidy, the gardens of everyday, working folk (aka peasants) exhibited a more free-spirited, practically minded ethos. Their gardens had a much more "higgledy-piggledy" planting of crops and flowers and weeds, many of which were important for both culinary and medicinal uses in the home.

More Resilient Produce in the Vegetable Garden

Perennial vegetables are incredibly useful plants and are another climate change must-have. Their longevity enables them to stand firm against more challenging conditions. With their deeper root structure, they seek out moisture more effectively during a dry period and absorb an excess of rainwater during storms, to the benefit of the plants around them.

The fact that the soil in which they grow remains undisturbed also provides a number of exciting benefits that we don't yet fully understand. It is the amazing underground world, with its complex fungal and microbial activity, that is so essential for the health and vitality of what's living above the ground.

Climate Change Garden Vegetables	Description
Asparagus (*Asparagus officinalis*)	Easily grown from seed or planted on from bought-in crowns, this delicious perennial will, when established, keep on providing for many more seasons. Be patient until you can take your first harvest in the third year and you will be paid back with plentiful returns for many years thereafter. These plants require a nutrient-rich, weed-free patch of ground as their long-term home.
Cardoon (*Cynara cardunculus*) and **globe artichoke** (*Cynara cardunculus* 'Scolymus Group')	These thistle-like, highly ornamental plants reliably return each year with just a little care if you live in a warmer climate. In cold climates, they are admittedly more challenging. In the case of globe artichoke, the heads themselves are edible, while with cardoon (also known as thistle artichoke) it's the blanching stalks that are highly prized in the kitchen.
Jerusalem artichoke (*Helianthus tuberosus*)	This rich, nutty tuber fits into the climate change garden well. Its tall-growing stems provide a useful barrier in the vegetable patch against wind to the benefit of your plot overall. Jerusalem artichoke plants cast shade, so that's worth bearing in mind, but otherwise these tasty tubers can be harvested in late autumn/early winter onward. You don't need to worry about leaving some in the ground to grow the following season because it will happen whether you mean it to or not! These plants have a strong survival instinct and produce tubers deep in the ground. New plants can always be expected to emerge with gusto in the spring.
Oca (*Oxalis tuberosa*)	These highly colorful small South American tubers are becoming more readily available for planting, thanks in large part to certain seed and plant suppliers and their work championing the growing of this nutritious plant.
Perennial kale (*Brassica oleracea* 'Ramosa Group')	These plants, also known as cottager's kale, would have been common in vegetable gardens in the past. They are damaged by caterpillars in the same way as other brassicas, but being perennial, they have the resilience to bounce back as if nothing has happened.
Rhubarb (*Rheum × hybridum*)	It may not be glamorous, but this stalwart "fruit" is very weather hardy and reliable. Once fully established, it will reappear like clockwork each spring, year after year, and can handle regular spring pickings. Once a rhubarb plant gets really big, its crown will need dividing (cutting in half with a spade), providing you with another rhubarb plant (or more!) to be planted out elsewhere in your garden.

Climate Change Garden Vegetables	Description
Sea kale (*Crambe maritima*)	Found wild on coastal rocky beaches, sea kale is traditionally forced in winter for its edible white stems, but the leaves and flower spikes can be harvested too.
Skirret (*Sium sisarum*)	This root crop was popular in the Middle Ages across Europe for its thin roots that taste of carrots and parsnip. It can be harvested all year.
Sorrel (*Rumex* spp.)	This zesty leaf provides a valuable addition to any veggie patch. Its leaves are delicious in any number of dishes, and it requires little to no care whatsoever.
Walking/tree onions (*Allium cepa* 'Proliferum Group')	This variety of onion will (as the name suggests) gradually "walk" around your veggie patch as it produces bulbs on the end of its foliage, which eventually bend over and plant themselves in the surrounding soil.

Herbs: A Place in Thyme

Herbs have a valuable role to play in the environmentally focused garden, attracting pollinators and beneficial insects, as well as furnishing us with very healthy, flavorsome material for the kitchen. Of the many herb varieties there are to grow, the perennials make an especially useful addition. In addition to being incredibly low maintenance, they help protect the soil against erosion over winter, providing structure as their roots bind the soil particles together for better absorption of rain.

Mixed Plantings and Rotation

Look at most home vegetable plots or community gardens and you will see patch upon patch of block- or row-planted vegetables. This is what we've been led to believe growing fruit and vegetables should look like. Similarly, the mentality of "grow it, harvest it, then pull it out" has been ingrained in much traditional gardening advice over the years. It's all very much built around this idea of keeping nature in check, with produce positioned in neat, weed-free lines for the designated growing season.

Although in an organic system the process of crop rotation is recommended (to ensure that the soil isn't drained of vital nutrients at the end of each season and disease does not build up), it doesn't have to be this way all the time. Growing different varieties of crop in an ordered succession (with brassicas

Sally's perennial bed with (left to right) sweet cicely (*Myrrhis odorata*), Chinese artichoke (*Stachys affinis*), sea kale (*Crambe maritima*), and Good King Henry (*Blitum bonus-henricus*).

following legumes, following potatoes, and so forth) will ensure order and overall soil health and fertility. However, if you plant different types of vegetables and even fruits all mixed in together in your bed, then the soil doesn't get depleted in the same way it would with monoculture, block-style planting. Instead, because you have fast-growing crops, such as lettuce, growing alongside a hungrier tomato plant, with a few Mediterranean herbs, pea plants, cabbages, and lots of nasturtium, carrots, some onions, and maybe some radishes, there is a natural biodiversity that enables you to ignore the crop rotation system entirely.

The key is to ensure that there is sufficient space between hungrier plants and balance between the crops that are planted together. Don't have too many of one type of crop in a particular area to ensure the soil remains fertile and pest buildup doesn't become an issue.

In fact, this more free-spirited method of growing can also help keep pest numbers in check. If you consider that it's often recommended to companion plant calendula or onions around carrots to help prevent carrot root fly attack, just imagine this principle on a much larger scale. It's essentially harder for a particular pest problem to get out of control when there isn't such a huge source of food in one place for them to get excited about. Instead, they have to hunt harder to find their target of opportunity. This provides another check in the resilience box.

These techniques become increasingly important when you consider that, with our changing climate, we are going to have to deal with a much broader range of pests and diseases than ever before.

Lowering Your Carbon Footprint: Working with Waste

In this rather crazy, materialistic world in which we live, and in which our reliance on disposable materials, especially plastics, has become a blight on the natural environment, finding productive ways to use such waste creates brownie points all round. It also helps to build confidence and develop the skills necessary to think on your feet, get creative, and come up with ideas and solutions regarding how to reuse readily available materials on your own plot. Here are some ideas to get you started.

Using Plastic Bottles as Cloches

Don't throw away plastic bottles. By cutting the bottom off a plastic bottle, the remaining plastic can be used as a protective cover for potted seedlings. Keep the lid off to enable air to circulate. The well-insulated young plant is able to thrive and grow in its plastic bubble. Such cut-down bottles can also be stored away at the end of the growing season and used over and over again, before eventually being added to the recycling bin.

Planting Trays and Pots

So many plastic pots from newly bought plants end up being ditched, yet they can often be successfully used repeatedly for many years to come. If we can abandon our "use once and throw away" mentality and treat plastics that have already been created as a potentially useful item with a long-term purpose, then that will be an important part of the solution. Of course, we should still look at more environmentally friendly alternatives. In the same vein, plastic fruit and vegetable containers and yogurt or food containers from the kitchen can be reused for planting seedlings. Even cardboard toilet roll tubes can be used as a biodegradable seedling pot.

Cardboard and Kitchen Waste into the Compost

Compost plays a vital role in ensuring the continued fertility of your garden. A huge amount of kitchen waste, including cardboard, can be successfully composted, dramatically reducing the amount of waste that goes to landfills.

Old Carpets

An old wool carpet is perfect for covering soil in winter and killing weeds.

Getting Inventive with Salvage

It's staggering and appalling in equal measure how much waste we send to landfills each year. If we can start to see these materials, which are so casually disposed of, as a valuable resource for the future, the less disposable our world as a whole will become. It's really easy and incredibly rewarding to find a valuable use for the sorts of items that would most frequently end up in a landfill.

Getting inventive with herb planting over winter.

An old carpet protects the soil through winter.

Old guttering repurposed as the perfect seed tray.

The Unexpected Perennial

Unless you live in a very cold climate, you can easily experiment with growing some of your plants longer than you usually do. The benefits of doing so are quite significant. In the case of the brassica family, for example, plants like kale and broccoli can be grown on for at least a few seasons. They will attempt to flower and set seed in their second year, so you just need to cut them back carefully when they do so. But the amount of produce each large, mature plant will provide is astonishing when compared to a first season's growth. Growing such produce in this way enables you to access a bumper harvest for little to no work at all.

Tips for Vegetable Growing in a Hot Summer

Growing vegetables in a dry, hot summer can be challenging, especially with heavy plant watering requirements. In addition to the amount of time involved in keeping produce sufficiently quenched, in places where water is rationed, this can work out to be expensive when rain barrel or storage tank supplies run out. But with careful planning, you can grow a wide range of vegetables successfully.

Mulch: Mulching helps your soil retain as much moisture as possible, and if you combine it with a no-dig approach (see page 74) and raised or mounded beds (see page 79), you will be well on the way to success.

Shade: Intense sun can damage young plants, for there are only so many cycles of wilting and rehydrating that leaves can take without dying. In addition to mulching, design your veggie plot to make the best use of natural shade and make sure you have enough space to erect shade structures. Taller crops can be planted to cast shade over more susceptible ones. For example, a row of sunflowers casts a light dappled shade and also reduces wind speed. Often gardeners pop a mesh cover over brassicas to stop butterfly and bird damage. In addition to protecting the crop, it casts shade, helping to reduce the temperature overall.

Sow and plant early: Spring sowings and plantings are likely to get established before the onset of hot weather and the drying out of soil. Later sowings and plantings are more likely to fail due to heat, so the ground needs to be watered well, although a really late sowing may be able to grow on through autumn.

Salad crops: Their leafy shoots, with high water content and shallow roots, mean that these crops are more vulnerable to water stress and need regular watering. Grow them in a small space and pick early. Lettuce, though, does bolt if stressed.

Grow more early crops: If dry summers become the norm, then think about crops that complete their growth before the heat of summer sets in, such as overwintered broad beans, onion sets, and early peas. Preserve the harvest for use later in the summer.

Grow lettuce between other crops and pick early.

Sprawling nasturtiums shade the soil.

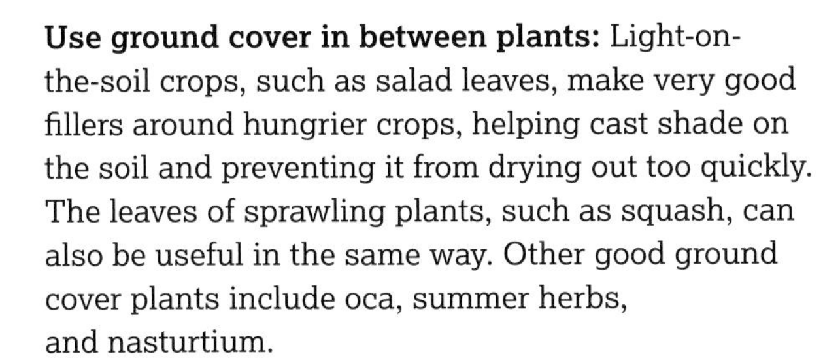

Use ground cover in between plants: Light-on-the-soil crops, such as salad leaves, make very good fillers around hungrier crops, helping cast shade on the soil and preventing it from drying out too quickly. The leaves of sprawling plants, such as squash, can also be useful in the same way. Other good ground cover plants include oca, summer herbs, and nasturtium.

Getting seeds to germinate: Ensuring successful germination in warm weather is tricky, especially with lettuce, so water the planting site several times before you sow to raise the moisture level of the soil. The water cools the soil too. Sow the seeds, cover lightly with soil, and don't water again until germination has taken place. It can help to rig up some shade to protect the seedlings from the sun's heat and to reduce evaporation from the soil.

Beets can cope with dry summers.

Brassicas vary in their tolerance to drought.

Vegetable Crops in the Climate Change Garden

Some common garden vegetables are adaptable and resilient to climate change, while others need special care. With a little knowledge, you can successfully grow the following.

Beets: These plants have surprisingly deep roots. They can be sown early and late to take advantage of the longer growing season. They don't need watering, as it encourages leafy growth, but the soil must not be bone-dry. Keep thinning the last sowing, so the last few have plenty of space to expand. Pull soil up around the roots to protect from frost and snow.

Brassicas: Modern varieties tend to have shallow roots, so a dwarf variety means dwarf roots too. In general, opt for the larger, traditional varieties that produce deeper roots and mulch well. Many of the brassicas, especially kales, can be started early when the soil has plenty of moisture, so their roots are established before a summer drought. Kale and Brussels sprouts don't need much water unless there is a long period of drought. Without water, their growth will be slow and the plants can look quite sad by the end of summer, but they pick up when the rain returns and will grow on through autumn into winter. Green sprouting broccoli (calabrese) needs moisture to produce flower shoots and stay tender, so is least suited to a dry summer. Instead, opt for early varieties or an alternative, such as the tasty broccoli rabe, or overwintering sprouting broccoli.

Carrot: Juicy varieties such as Nantes have less fibrous roots, which means the roots are more likely to split early in dry conditions, but the advice is not to water, as it lowers yield. Variety selection will become increasingly important.

Celery and celeriac: Both these crops are susceptible to drought and will need watering.

Chard: Despite its large leaf area, chard is pretty tolerant of dry conditions and can be likened to beets in its water needs.

Jerusalem artichoke: This crop has deep roots, so there is no need to water it. In fact, too much water can encourage leaf growth rather than tubers.

Presprouting Seeds

Presprouting seeds is a technique used by growers in dry areas to save water and ensure successful germination. Simply soak seeds overnight, drain them, then leave in a dish. Rinse and drain once or twice a day until the first roots appear, at which point they are ready for sowing. To make it easier to sow them, use a fluid planting method. This involves mixing the seeds into a starch gel to help disperse the seeds down the row. It's simply a couple of tablespoons of cornstarch mixed into a pint of boiling water, which is left to cool and form a gel. The seeds are mixed gently into the gel and poured into a plastic bag. When you are ready to sow the seeds, cut off the corner of the bag (like an icing bag) to create a small hole through which to squeeze the gel into the planting site (it's wise to practice this with old seed first). Using this method ensures the seeds are surrounded by a gel and are not water stressed. Plants will appear quickly too, as they are pregerminated.

The leafy growth of leeks shades the soil.

LEARNING FROM THE PAST: **Heritage Vegetables**

Kitchen gardens of the past lacked hoses, and gardeners had to rely on saving water in the garden or lugging it by bucket. Heritage vegetables may be the answer in the future. Look for slower growing, deep-rooted varieties. They may not yield as much but will be more resilient to a hot, dry summer. For example, heritage runner beans have much deeper roots than the modern varieties selected for high yields and rapid time to harvest.

Leek: Many gardeners have found that leeks survive dry conditions well, but they do need a moisture-retentive soil. They grow slowly in warm weather but pick up again in autumn. It may be wise to opt for vigorous varieties that mature later in the year when there is more rain.

Legumes: In drought conditions, runner beans suffer more than French beans, and more exotic choices such as chickpeas, lentils, and butterbeans thrive. Broad beans and fava beans can be sown in autumn to get an early start and complete their life cycle before the summer heat. Early varieties of peas may come into their own, as they too can complete their growth and fruiting before the dry weather kicks in, but second and third sowings may have lower yields in dry weather. With runner beans, make sure they have plenty of organic matter in the soil. Watering is particularly important at the start of flowering and during pod formation.

Lettuce: This is a cool weather crop, and many varieties go to seed in hot, dry summers. Also, they fail to germinate when temperatures rise above 70°F (21°C). A leafy type may be more resilient, especially oakleaf lettuce, which has proved to be very resilient to high temperatures and lack of water, plus it can survive light frosts. Reduce the heat burden on lettuce by planting it in the shadow of taller plants or by shading the bed.

RIGHT: **Some varieties of lettuce are more drought resilient than others.**

Onion: These are established in spring, so there should be moisture in the ground, but a dry spring means water will be needed for good establishment. Don't water after midsummer, as this delays ripening.

Parsnip: This is a deep-rooted crop, so there is no need to water.

Potatoes: Coming from the Andes, potatoes are more tolerant of dry summers than many other vegetables. There is a wide choice of varieties, but the more resilient types are those that are very late and slow maturing and that can be lifted in autumn after they have time to recover from any drought. The advantage of a drier summer is less blight. It's vital to ensure sufficient water during the early stages of tuber initiation and around flowering.

Radicchio: This originates from the alpine regions of northern Italy and is hardy. It is sown later in the year and harvested throughout winter. In fact, it needs cold for its leaves to turn red, so a mild winter may produce less colored radicchio. With this crop it's important not to sow too early, otherwise you end up with a green lettuce that's hairy and bitter, so make sure you sow in the heat of summer and let it grow through the cold weather.

Rutabaga and turnip: These have deep roots, like beets, and can continue to grow through drought. Plants from a late sowing will struggle through a dry summer and appear stunted, but they have the capacity to recover and grow into decent-size roots by early winter.

Spinach: This leafy plant needs watering, but it can survive wet winter weather.

Summer squash and winter squash: It's best to sow these in small groups, as they need the right soil temperature to germinate. To get good growth, plant on a mound of compost to provide moisture retention and fertility. They wilt in hot weather, with some varieties being more prone to heat than others, but they recover. However, wilting may indicate that they are planted too close together and are not getting enough water from the available root zone. There are some newer compact varieties, but their weaker root systems mean they are less able to cope with drought. Summer squash, such as yellow crookneck, are particularly drought tolerant and high yielding, as their large leaves help shade the soil and keep temperatures down, but green zucchini need a rich soil with plenty of moisture-holding capacity and require regular watering. You could start them earlier, planting them out under a cloche for frost protection.

Squash leaves shade the ground.

Sweet corn plants have a poor root system.

Sweet Tomatoes

Ethnobotanist James Wong recommends spraying a dilute solution of soluble aspirin (half a tablet in 1 quart [1 L] of water) over your tomato plants to boost the sugar content of the fruits. Aspirin resembles salicylic acid, a plant stress hormone, so it tricks the plant into thinking it's under stress and it sends more sugar to its fruits, and you get sweeter tomatoes. But just as importantly, it boosts the plant's resistance to cold, drought, heat stress, and late blight. Try it next time! Another plus is that tomatoes grown with less water tend to be tastier, as there is less watering down of flavor.

Sweet corn: This water-demanding crop often has poorly developed roots and requires regular watering. You could opt for some of the more drought-hardy field corns, but choose your variety carefully, as many are bred for making flour and other uses.

Tomato: This crop loves the heat and a longer growing season suits them well. When planting, trim the lower leaves and place the stem deep in the ground so that leaf scars are covered by the soil and will sprout roots.

If you have plenty of space, you may want to think about growing tomatoes without support, in their natural form as sprawling vines that cover the ground. This habit shades the roots, allows the stems to root, and draws nutrients from a further point in the ground, but harvesting fruits is not so easy.

Especially drought-resistant varieties include 'Roma', 'San Marzano', 'Black Krim', 'Early Girl' (cool tolerant, so this one gets an early start), 'Sweet Million', 'Sungold', and other cherry varieties. The darker pigmentation of the chocolate and black tomatoes blocks the sun's rays rather like a sunscreen, preventing sun scald and reducing water demand. 'Principe Borghese' can be sown early for a crop in early summer and then sown again for an autumn crop.

Growing tomatoes under cover in hot weather can be problematic due to the extreme heat and high humidity experienced in polytunnels during the height of summer. This can adversely affect pollination, which, combined with fewer pollinators flying to assist the pollination of tomato flowers (which are self-fertile but are boosted by pollinators), may mean a poor harvest. For a polytunnel crop, choose varieties that flower early and reach picking stage sooner, and look for disease resistance, because late blight spreads more readily in hot and humid conditions.

Some Vegetables to Try

As our summers get warmer and the growing season lengthens, there are a number of more unusual vegetables well worth trying.

Butter beans or lima beans (*Phaseolus lunatus*): Runner beans suffer in a dry summer, so you might want to try growing butter beans instead. Grown in the same way, they positively enjoy the heat. They produce short pods with two or three fat white seeds, which can be dried and used in soups.

Chickpeas (*Cicer arietinum*): This legume needs a long season. It's frost sensitive, so the young plants are transplanted after the last frosts or protected by cloche or a row cover. The pods contain one or two seeds that can be harvested green or left to dry.

Cowpeas or black-eyed peas (*Vigna unguiculata*): These are grown in the southern United States, Africa, and Asia. It's an easy-to-grow, warm weather crop. The peas will tolerate heat and drought. The pods are picked and boiled or stir-fried.

Chickpeas love warm, dry summers.

Edamame or soybeans (*Glycine max*): This bush bean needs a long growing season. Plant the seedlings in a well-drained soil in a sunny position either after the risk of frost has passed or earlier under a row cover. If you want edamame beans, pick the pods while they are still green. The easiest way to remove the beans from the pod is to boil them so the beans slip out. If you want soybeans, leave the pods on the plant to ripen and turn brown.

Huauzontle (*Chenopodium nuttalliae*): This tall plant is a relative of quinoa. The edible leaves are green with pink stripes. It's drought resistant and the pink seeds can be saved and used to make tortillas. It grows fast, so can be sown in late summer.

Okra (*Abelmoschus esculentus*): This heat-loving crop is originally from Africa, but is now grown across the world. It comes in lots of different varieties, so you need to find one that will suit your conditions.

Tomatillo (*Physalis ixocarpa*): This is the savory version of the cape gooseberry, and it can be used in salsas and sauces. It scrambles over the ground and is slow to mature. As the fruit ripens, the husk peels back to reveal a green- or violet-colored fruit. They need a long summer to mature but can produce prolific harvests.

Yardlong beans (*Vigna sesquipedalis*): These are tropical beans and, as the name suggests, the plant produces long, thin beans. It's usually grown in a polytunnel but will do well outside planted along a south-facing wall.

Edamame beans are an easy-to-grow crop.

Growing Under Cover

Having the ability to cover your plants, should the need arise, is invaluable in the face of volatile weather, especially early and late in the year. Cover will provide protection for your more fragile seedlings and enable those frost-sensitive plants to grow for a few weeks longer. It enables you to tap into a much longer growing season, as well as providing a safety net against the worst of the extreme weather of the future.

There are many ways to go about providing cover, from the installation of a polytunnel (hoop house) or greenhouse to the use of protective covers and structures, which provide many of the same benefits, just on a smaller scale. No matter your budget or site, there are a range of solutions that can assist your food-growing efforts.

RIGHT: **Kim picks produce all year-round in her polytunnel.**

Polytunnel Growing

These useful covered structures are increasing in popularity because they enable the "grow your own" enthusiast to greatly extend the range of edibles they can grow in new and exciting ways. From reliable crops of tomatoes, cucumbers, eggplants, and peppers to sweet potatoes, melons, and grapes, all these fruits and vegetables thrive under cover. In addition, tunnels provide an undercover outside space that can be used by the gardener on even the wettest of days, so it will double as an attractive and protective haven in which to sit and ponder your climate change gardening days ahead.

Polytunnels are a lot cheaper to buy than greenhouses and relatively straightforward (with a little help from a few friends) to put up. If you can find the space, they really take crop-growing to a new level and allow you lots of room for experimentation. They let you widen the range of fruits and vegetables that can be grown, will overwinter many more plants successfully, and help you save seed from a greater diversity of plants than would be possible in an outdoor space.

RIGHT: **The longer season provided by a polytunnel favors crops such as tomatoes.**

WHERE TO LOCATE YOUR TUNNEL

Begin by checking local building codes to see whether you can build and use a polytunnel where you live. Here in the United Kingdom, generally speaking, planning permission is not required for a polytunnel if the structure is less than 3 feet (1 m) tall and doesn't take up more than half of your garden. It also shouldn't be positioned too close to a road, or in such a way as to cause offense to a neighbor by blocking light or being too visible from their garden.

Increasingly, community gardens are becoming more receptive to the idea, although most tend to favor structures on the much smaller end of the scale. But wherever you live, it's wise to check first with your local planning authority.

From an installation perspective, find a sunny spot that is as level as possible. Position the polytunnel with a north-to-south orientation because it enables you to get the most sun possible during the day. However, with more periods of extreme sunshine on the horizon, it's worth thinking about a position that affords some shade for part of the day. East-to-west positioning, therefore, may be more suitable.

Nearby hedging or trees would be beneficial in terms of early morning or late afternoon shading, so the tunnel doesn't heat up quite as much in the peak of summer. Hedging can also reduce wind speeds (see chapter 3) and help screen your polytunnel, enabling it to blend better into its surroundings. Trees and shrubs also help soak up excess water running off the polytunnel and minimize waterlogging in and around your plot, making a valuable contribution to your climate change garden.

MAKING YOUR PLASTIC LAST LONGER

The polytunnel structure has the potential to last a very long time (decades, in fact). It's the plastic cover that requires replacement more often. Manufacturers recommend every four years because (over time) UV light causes the plastic to become brittle and less visible light can pass through. The tunnel becomes less and less effective as a growing space. That said, one of Kim's tunnels is still going strong after fourteen years with the same plastic cover, so when your structure is well cared for, the plastic may last much, much longer than the recommended guidelines.

Keeping the plastic clean and free of algae, fungi, and debris, such as fallen leaves, is very important, as such materials will erode the plastic over time. Cleaning the plastic and keeping the tunnel clear is a vital job in the polytunnel gardening calendar.

LEFT: **Spring in Kim's polytunnel.**

VENTILATION

If you ever hear someone referring to polytunnel growing as being hard or tricky (which it absolutely isn't), airflow is probably the main reason why. Ensuring sufficient airflow is incredibly important to the health and vitality of the ecosystem within your tunnel.

It's obvious that during the potentially long, hot summers, doors will be opened more often to help keep the daytime temperature at a healthy level and to provide airflow through the tunnel. Airflow is just as important in winter, so once or twice a week, try to open the doors to allow a natural breeze to flow through. Not only does this help prevent air from becoming stagnant, but it is also critical to reducing the threat of fungal disease building up on your winter salad leaves.

A handheld hose nozzle is effective but time-consuming.

WATERING

In a structure that is impervious to rain, plants will require access to this vital liquid of life. However, this doesn't mean the process of watering need be laborious or overly time-consuming, even during a heat wave (as explained on page 148). Instead, it's about setting up systems to minimize the time and effort involved in keeping your produce sufficiently quenched all year round.

Sprinkler systems: Though they are often frowned upon because they can encourage blight on tomato plants (with moisture landing on the leaves), this system works really well for much of the year when employed in the morning, with all doors and windows opened wide. If using this method of watering, a mixed planting system (as outlined on page 128) is also preferable because it ensures a greater biodiversity from the ground up, affording greater resilience to fungal buildup in the process. A sprinkler can be run either from a main water supply or, with the aid of water pump, it can circulate rainwater from your storage tank.

On the ground drip watering: To enable a slow flow of water direct to the roots of your plants, you can use a drip watering system. It might be a simple old hose with some holes in it connected to your rain barrel, or it could be a sophisticated timer-controlled system with a manifold feeding several hoses with drip notches, or anything in between.

Handheld hose nozzle or watering can: At the more time-consuming end of the watering spectrum, this can still work well in smaller tunnels as a viable watering option. A hose with a sprinkler attachment is much less effort than a watering can (unless you like spending your days going back and forth to collect water). Again, if you're keen to use recycled water, then a pump will enable you to distribute your rain barrel supply with ease.

AVOIDING THE BUILDUP OF PESTS

It is a relatively common assumption that, inside a polytunnel, you are much more likely to experience a buildup of pests such as aphids. However, this is simply not the case if you employ a more biodiverse planting system with much more mix and match than block planting, encourage wildlife in, clean your tunnel, and allow sufficient airflow. There is more on this in chapter 6, but wildlife, such as birds, ladybugs, ground beetles, amphibians, and so on, all have a vital role to play. In a biodiverse tunnel, it's much harder for one form of pest to get out of control, as the problem pest is always food for something else, and therefore simply becomes a food source, rather than a detriment to your plants and produce.

RIGHT: **Blackflies can be problematic in a polytunnel.**

DEALING WITH EXTREME HEAT

Yes, a polytunnel will get very hot indeed in the midst of summer, especially if the doors and ventilation panels are all shut. It really doesn't take long for the sun to raise the temperature inside to an unbearable degree. This is why planning ahead to prevent such heat buildup is key.

During the warmer months of the year, it pays to keep everything you can open to the elements to encourage a cooling airflow as much as possible. Erring on the side of caution, in favor of a cooler tunnel overall, saves you from worrying about your plants wilting when (for example) a cooler summer morning turns into an absolute scorcher when you are at work and unable to do anything about your plants. If you're worried about rabbits or other wildlife with plant-nibbling tendencies getting in overnight, then simple barriers can easily be constructed or used to provide protection.

There is a wealth of advice on dealing with extremes of heat in chapter 2, much of which applies to protecting plants, both inside and out. Suffice to say, the techniques outlined were tested to the limit (successfully, we are pleased to add) during a recent heat wave, when Kim's private water supply ran almost dry. This meant that, with plentiful rain barrel reserves but a distant memory, polytunnel planting had to deal with being given a deep-soak watering just once a week. The ground cover, mulching, and mixed planting enabled most of her plants inside to thrive surprisingly well despite the more challenging conditions, which just goes to show how much lower maintenance summer watering can in fact be with the right soil protection in place.

Another option for the polytunnel is the addition of mesh or vents on the sides or in the roof to provide greater airflow and cooling during periods of extreme heat.

HANDLING STRONG WINDS

A well-installed polytunnel will be able to stand firm against severe winds, as long as doors and all openings are kept securely shut. If you know extreme winds are forecast, it's also important to err on the side of caution and remove any potential flying debris from the vicinity, such as lightweight garden furniture or buckets, to avoid any damage.

Greenhouses

These valuable glass structures are also incredibly useful to have in the climate change garden. While many of the same advice regarding polytunnels also applies, these structures can actually get even hotter when located in a south-facing position, so that is something to bear in mind.

In our experience, aphid buildup can be more of an issue in greenhouses and so, as much as possible, you want to try and encourage beneficial wildlife such as ladybugs into your under-glass haven.

Planting in the ground is also preferable, where possible, to planting in containers, as these are often small or shallow. As a result, they can dry out rather quickly, which is far from ideal in such a torrid environment.

Seed Saving

While saving all of your own seed from the vegetable garden each year would be an extremely time-consuming and, frankly, laborious thing to do, working with at least a few crops in this way is viable. It is also extremely beneficial in the battle against the metrological volatility of the future because the seed that you save (when done correctly) will have been genetically adapted to the growing conditions of your individual locality. This is especially the case when seeds are saved time and time again over a period of years, with the best specimen plants grown on to flower and set seed for storage.

Seed saving used to be a firm part of the gardening calendar. It's only in more recent years that we've become so accustomed to buying seed every year. Although flicking through seed catalogs in the murky depths of winter (or indeed any time of the year) is undoubtedly an exciting and wholesome pastime, saving a little seed is also an increasingly important part of the climate change gardener's routine.

Seed saving is incredibly easy to do, and you can save seed from more types of fruit and vegetables than you might otherwise think. The climate change credentials of saving some of your own seed are very strong indeed. No matter where you live and what you like to grow, there will be some crops that are easy to work with, producing precious seed adapted to your individual grow-your-own space.

Saving seed from the broad bean is easy.

Once you understand the basics and get some practice under your belt, the potential to save from a much wider variety of crops is opened up to you. It's possible to save from most of the vegetables that you grow on your vegetable patch, though sweet corn is a particular exception. It requires a much wider genetic source base (that is, it requires many plants to be successful). This makes it unrealistic to save sweet corn seed at home, because you wouldn't have room for hundreds of plants, and what on earth would you do with all that seed, even if you did?

If you have never saved seed before, have a look at the following table that lists some of the easiest crops from which to get started.

Easy Crops	Method of Seed Saving
Peas and French beans	With these self-pollinating plants, simply leave a few pods in place to fatten, yellow, and dry before collecting. Or harvest the plump peas at the end of the season and leave the pods near a radiator or in an open cupboard to dry out before storing away until the following year.
Lettuce and arugula	Ensure that your lettuce has a long enough season to flower and set seed for you to harvest and collect. The only tricky part can be in a wetter summer, when seeds don't have the opportunity to dry in situ and go moldy. In such weather, you'd be better off working with plants moved inside.
Tomatoes	Harvest your tomato of choice and scoop out the seeds. You can then either leave them to dry on a piece of paper for planting the following season or, for a slightly fussier method, drop the seeds into a small jar of water. The jelly-like coating rots away, leaving just the seed, which can then be left out to dry.
Chili peppers	If you are growing a few varieties of chili peppers, they can cross-pollinate, and the resulting offspring plants may not grow "true to type." This isn't a bad thing necessarily, as you will have just created your own homegrown variety, which could be milder or hotter than the parent plants.

Tips for Success

Avoid FI varieties: It's best to save seed from open-pollinated varieties. It's easy to do—just look for seed packets that say F1 on the cover, and don't save from these plants. The reason you can't use these plants is that they have been created from a cross between two carefully selected parent plants and the resulting seed could end up with the characteristics of either and won't grow "true to type." You just won't know what you'll end up with. Instead, you want to save from varieties that can adapt to your own growing space and that display the genetic resilience, taste, and produce qualities that you are looking for.

Pick the best-looking specimens from which to work: If you think about the fact that you are going to be growing the offspring of the parent plants from which you chose to save seed, then it makes sense to select the healthiest, best examples of that parent crop.

Avoid early bolting plants: In the same way that you want to choose the healthiest specimens, you don't want to carry on less attractive plant traits, like early bolting. Avoid saving seed from any arugula, radish, lettuce, or other types of vegetables that bolt early. Wait and work with the specimens that don't have this unattractive habit.

Leave your carrots in the ground over winter and they will flower in summer.

Nurture resilience: While it's obvious that you wouldn't wish to save seeds from plants that have been weakened by disease or pests, try to pick the plants that have shown hardy qualities in the face of adversity. So, if some of your pea plants have thrived better than others during the course of a heat wave, save seeds from them. These are exactly the genetic qualities you want to be selecting and nurturing to help your plants ride out the more extreme weather ahead. Seed saving is such an important pastime for many reasons, but this is perhaps the most important element when it comes to climate change gardening. Backyard seed savers can be the guardians of a more resilient homegrown future.

Unfortunately, seed from the brassica family, carrots, parsnips, leeks, and other species cannot be saved reliably because they easily cross-pollinate and you need a lot of plants for enough pollination to occur to produce seed. Cross-pollinating crops has the potential to cross with other species from the same family, as they are insect-pollinated. For example, if you have purple-sprouting broccoli flowering at the same time as kale and a bee has landed on both in quick succession, then it's likely that pollen has been transferred between the two and you risk ending up with a cross. The resulting plants grown from the saved seed might prove to be okay, but it's a risk and you could end up growing a crop that turns out to be terrible in terms of taste and texture, so is best avoided.

Advanced Crops	Method of Seed Saving
Beets, spinach, Swiss chard	These are biennial crops. Either leave 20 plants in the ground over winter and they will flower in the spring, or in the case of beets, harvest the roots in autumn and keep in storage. Reserve the 20 best specimens and plant them back outdoors in the late winter/early spring for growing on. These crops easily cross-pollinate, so if they are grown within several hundred feet of each other, they will outbreed. Do not attempt to save seed from non-isolated plants unless you really like surprises.
Carrot	Biennial crop that flowers in its second year. This plant produces the most delightfully ornamental flowers. Leave 20 plants in the ground to flower the following spring, making sure you only work with roots that have desirable characteristics. For example, a pale yellow carrot (when it should be a vibrant orange) will add those traits into the genetic mix of the seed. Garden carrots can cross with wild carrot relatives, including Queen Anne's lace, so for this reason, it's better to grow carrots you plan to save seeds of undercover, so wild pollen cannot make its way to the blooms.
Kale	Most brassicas are biennials, which means they will flower and set seed in their second year, so they need to remain in situ during this time. The key is to ensure that when they do flower, there isn't anything else also flowering from the brassica family at the same time. Broccoli is normally the troublemaker is this regard, so it's the main plant to watch out for and cut back if it does attempt to flower at the same time. A lot of seed will be produced from 20 plants, but kale seed stores reliably for a number of years, so your time and space investment will be paid tenfold in return.
Leek	As with carrots, this crop can be grown (and left in the ground) until the following spring in a relatively small space. To ensure genetic strength and vitality, and to create seed that can be grown on and bred from, a minimum of at least 20 plants is required to ensure quality and genetic variation. In reality, specialist seed-saving institutes and heritage seed libraries will use far more plants, but for the home-based enthusiast, it's possible to get away with a smaller number.

Calendula will readily self-seed.

Allow a few arugula plants to flower and set seed.

The Best Self-Seeding Plants

Along with the process of collecting and storing seed over winter to be sown in the spring, there are plants that are quite capable (and possibly better off) doing it naturally for themselves. These self-seeders include coriander, parsley, nasturtium, arugula, calendula, and many others.

Sometimes you don't need to do anything at all. Just wait for lots of lovely resilient seedlings to burst into life in spring from seed that was dropped the previous growing season. All you will need to do is to gently lift and transplant them elsewhere around your vegetable garden.

IN THE ORCHARD

We love our fruit trees, and for many of us, a garden is not complete without a fruit tree or two. Yet, perhaps surprisingly, our precious trees and their fruits are under increased threat from a changing climate, loss of pollinators, and a rapidly multiplying number of pests and diseases.

Apples and many other top fruits are grown in temperate zones around the world, including South Africa and Australia, so being warmer in summer is not necessarily a bad thing. The problem facing them has more to do with mild winters and wet springs and a generally more erratic climate overall. What will happen in the future as our climate becomes ever more volatile?

When you consider that the fruit trees we choose to plant now will be reaching their best production in ten years' time and will continue to produce for many years thereafter, what should we be planting? Will our favorite heritage varieties, which were developed a hundred years ago or more, cope in a changing climate? Is it better to stick to local varieties and hope they are better positioned to adapt? Should we look to new varieties? Or should we think about using different rootstocks? Amidst all of these questions there are three key issues to consider: the chill factor, the rootstock, and pollination.

LEFT: Sally has a small orchard in her walled garden.

Apple Color

Surprisingly, the color of an apple depends on temperatures too. To get a lovely red color you need a large fluctuation between day and night temperatures, with the best color developing when there are some cold nights in autumn. If the apple crop is early and the fruits start maturing in summer, or high summer temperatures extend well into the fall, the likelihood is less intense color. High summer temperatures will also reduce fruit firmness, texture, and eating quality.

The Chill Factor

Mild winters are not good news for our fruit trees. Apples and other tree fruits need to experience a minimum number of chill hours (a period when the temperatures are below 45°F [7°C] during the tree's dormancy period). This cold period initiates flower bud formation. Too few hours and flower buds may open late or not develop at all. Each variety has its own minimum number of required chill hours, hence it's generally recommended that you grow local varieties that are adapted to your climate. For example, there is no point growing northern cold-climate varieties with a high number of required chill hours if you live in the south because the chill hours may not be achieved in a mild year. However, you can do the opposite and grow a low-chill-hour apple variety further north.

In general, many apples need 500 to 1,000 chill hours. Yet, as winters get milder, there will be fewer chill hours, so looking forward, it may be wise to select varieties that need less. Look for varieties that are recommended for warmer areas and may only require 300 to 500 chill hours. Already some commercial orchards are looking to plant more of the international varieties, such as Granny Smith, Golden Delicious, Gala, and Fuji. Other popular tree fruit species have chill requirements, too. Plums are likely to be particularly hard-hit, as they have the longest chill requirements (800 to 1,000 hours), while some peaches need as few as 100 hours.

The choice of rootstock will be important in the future.

Rootstocks

Most fruit trees are grafted. That means that the shoot system of one plant has been joined to the root system of another to create a new plant. The rootstock used determines the vigor of the tree, its ultimate size, fruit-producing ability, and often disease resistance, while the shoot determines the variety of fruit. There is a wide range of rootstocks for each fruit species.

For apples, the first rootstocks came from East Malling in Kent, England, more than a hundred years ago, and they were used around the world. Further breeding took place in different research institutes all over the globe. The East Malling rootstocks are identified with an M and EMLA. For example, M25 is for a traditional full standard tree growing to 16 feet (5 m) or more, M27 is a dwarfing rootstock that produces trees of just 5 feet (1.5 m) in height, and MM106/EMLA106 is for a tree up to 13 feet (4 m) in height and spread, which is ideal for small orchards and training into espaliers. There are also the G rootstocks from Geneva, New York, which are the most common in North America, many of which confer resistance to fire blight, a serious bacterial disease of apples and pears (see page 161).

Now breeders are looking for new rootstocks that may be more tolerant of the less predictable conditions that climate change brings and also more resistant to some of the diseases we may see in the future. For example, rather than choosing a tree with the popular MM106/EMLA106 rootstock, it may be better to go for the similar MM111/EMLA111, which produces a slightly taller tree. It's more vigorous and can cope with light and heavy soils, drought, and wet conditions.

When we think about climate change, we need to consider deep roots that can reach down and find water in a dry summer, so it may be better to opt for a larger tree than a dwarf one. Most fruit tree suppliers stick to a small range of tried-and-tested rootstocks, but if you look further afield, there may be trees on different rootstocks, which may be more suited to future conditions in your region.

Problems in the Mountains

In China, 100 percent hand pollination of fruit trees is common in remote mountain areas. Teams of people with paintbrushes work long hours to transfer pollen between flowers to ensure a crop of fruit. In fact, hand pollination is common across the Hindu Kush of Nepal, Pakistan, and India, where poor fruit set is experienced. The reason for this is the loss of habitat for pollinators. Hope is on the horizon, however, as farmers in the region are planting more native flowers to supply nectar to pollinators and bees are being encouraged where once they weren't.

Pollination isn't the only problem, though. The Himalayas are warming up quickly and winters are milder, so fruit trees are experiencing too few chill hours and their buds don't open. Growers are having to plant orchards higher in the mountains where it's colder. Apple orchards are now found at 10,000 feet (3,000 m) and above and they are running out of slopes. There are problems in Germany too, where it's traditional to plant fruit trees on alpine slopes to create shade for livestock and to have a crop. Yet with increasingly limited chill hours, farmers are now looking to grow traditional trees grafted on vigorous rootstock that produce huge crowns and deep roots, so they are more resilient. The trees may take ten years to fruit but are more likely to survive.

Pollination Woes

In temperate regions, apple trees are flowering an average of seventeen days earlier in spring compared with just sixty years ago. This risks an early flowering variety being hit by frost or there not being enough pollinators around when the blossoms open, which is a real worry for fruit growers. An early spring may also mean insufficient chill hours. However, there is another consideration—pollination groups.

Fruit trees are grouped according to their flowering period; group 1 (or group A) blooms in very early spring, while group 7 (group G) is in very late spring or early summer. Each group is in blossom for about two weeks, overlapping briefly with the group in front and behind. For example, a group 3 variety will probably come into flower in mid-spring and be pollinated by trees in groups 2, 3, and 4. Some varieties are self-fertile, so they do not need to be planted near another of the same pollination group for fruit, but for other varieties, you have to grow several different compatible varieties to ensure cross-pollination. This is where a healthy insect population comes in. If you only have a few trees, you want plenty of visiting pollinators that fly on to other fruit trees in the area to increase cross-pollination.

However, the shift in flowering time due to climate change is not happening in a uniform manner. It's not as simple as group 1 varieties flowering a few weeks early followed by group 2, and so on and so forth. Climate change is affecting varieties in a different way. For example, some group 4 varieties are flowering at the same time as group 1, while some group 3 trees have been unaffected by the changes and are flowering at their usual time. So, the practice of selecting varieties according to their pollination group is being turned upside down.

A bee brick provides a home for mason bees.

Honeybees versus Solitary Bees

Honeybees are not the only pollinators. There are many more pollinating insects, such as hoverflies and bumblebees, but the most effective pollinators are the solitary bees. Research has found that 600 solitary bees may be able to pollinate an apple orchard as effectively as two full hives of honeybees. Since honeybee hives have between 10,000 and 15,000 occupants, that means that 600 solitary bees could potentially do the work of 30,000 honeybees.

Solitary bees include the mason bee (*Osmia* spp.). These small, nonaggressive bees nest in cracks in walls and mortar and can be attracted to your garden through the provision of overwinter sites. You can even buy bee bricks specially made for mason bees. These bees also need an area of moist earth in spring so they can collect building materials for their nests.

To attract a diversity of bees, have a few wild areas in your garden and plenty of nectar-rich flowers. Native wildflowers are particularly important, as they are better at attracting bees than cultivated varieties. If you have the space, think about putting a beehive in your orchard. If you don't have time to look after bees yourself, see if a local beekeeper wants a safe home for one of their hives.

High temperatures have scorched the leaves of this apricot.

Stress of Extreme Weather

If we experience rainy winters, it's likely that soils will be wet for longer, resulting in slow drainage and root death caused by waterlogging as trees stand in water. Wet weather can cause problems for young trees in the first few years of their establishment, especially from crown rot caused by *Phytophthora*. Some rootstocks and varieties are more susceptible, the MM106 rootstock being the most vulnerable. Symptoms include poor growth, yellowing leaves, and underground, an orange-red rot.

Stormy weather may lead to more tree damage and uprooting, while hail can ruin an entire crop in just a few minutes. Unseasonably mild weather in winter will lead to weak flowers and poor fruit set, with a greater risk of frost damage.

There are problems associated with hot, dry summers, too. Fruit trees can survive a short drought, but a prolonged one causes root stress, defoliation, premature fruit fall, and small fruits. The trees can experience heat stress if temperatures exceed 68°F (20°C) for long periods of time, with sun-scorched leaves and damaged fruit. Lower rainfall over spring and summer will also reduce yield.

Wild Apples

We are fixated on apple varieties that are grafted onto a rootstock, but what about the wild apple that reliably grows from a seed? Wild apples are everywhere. Sally has several growing in hedgerows on her farm. The fruits are larger than a crab apple, and in her case, sweet. The fascinating thing about wild apples is that you don't know what you are going to get—a sweet eater, a tart-tasting cooker, or one that's perfect for cider making. People are now looking afresh at wild apples. In the past, the shoot of a good tree would be grafted onto a rootstock to create a new variety, but now it may be better to just sow a seed and see what emerges!

Our weather is going to be unpredictable, which in some regions will translate to more gray, rainy days. This will mean less photosynthesis, less sugar, and smaller fruits. The resulting fruits may drop early or have poor storage qualities. So, as you see, the threats on the horizon are multifarious and very real indeed.

Arrival of Pests and Diseases

Earlier springs and later autumns mean a much longer growing season, not just for our fruit trees, but for their pests and diseases as well. Mild winters may lead to more pests successfully overwintering and becoming active earlier in the year, while warm summers may see an influx of pest threats from other regions.

There may also be more defoliating pests around, and where conditions favor some of our minor pests and diseases, we may see them become major issues. For example, fire blight caused by the bacterium *Erwinia amylovora* is very problematic in some parts of the world. The bacteria overwinter in bark cankers, and then in spring the bacteria ooze out and infect the tree's inner bark by way of the blossom. At the moment, its spread is restricted by cold weather around the time of flowering. But as springs become warmer and more humid, conditions will favor the bacteria, which are spread by bees and rain.

Wet winters may see a surge in scab caused by *Venturia inequalis*, although it will be limited by hot, dry summers. Powdery mildew, too, may become a problem as it likes hot, dry summers. This common disease of apples and pears turns young shoots gray-white, distorting the leaves and causing the blossom to drop. The causative agent, a fungus, overwinters in the buds and then becomes active in spring.

Protect Your Trees: How to Provide Resilience

Soil around fruit trees growing in gardens and orchards is just as important as soil in the vegetable plot. Spread a 2- to 3-inch (5 to 7.5 cm) thick hardwood chip mulch around the trees to a diameter of 3 feet (1 m) in spring to retain moisture and protect the roots from temperature extremes. This controls the weeds too, which will compete with tree roots for water and nutrients. If you source wood chips rather than use your own, make sure the chips have not been taken from wood infected with honey fungus. There is some evidence that willow chip mulch helps combat apple scab, so that's one source to look out for.

Spread a thick layer of mulch around the tree but not right up to the trunk, as this can cause rot.

A medlar will boost diversity within an orchard.

A Diverse Orchard

As mentioned before, diversity is often the key to success. If you are planning a new orchard, however small, try to choose a wide range of varieties. This will make your orchard more resilient. A diverse orchard supports a wider range of insects and other animals too. If you are choosing apples, mix up the dessert, cider, and cooking varieties and their time of harvest, and choose different pollination groups to ensure successful cross-pollination. Late-flowering, late-maturing varieties generally need more chill hours, especially the cider varieties, so these are better suited to more northern sites. Think carefully about your choice of rootstock too. Also include a good mix of other fruits, such as pears, plums, peaches, and even more unusual fruits like medlar and mulberries, and maybe some nuts, too.

Think carefully about the siting of a new orchard. Try to avoid exposed windy sites or plant the fruit trees behind a windbreak. Spend time preparing the soil to make sure it drains well. A warm, sunny spot attracts more pollinators. You can even plant the odd crab apple to provide extra pollination opportunities. Varieties such as 'Golden Hornet', 'Winter Hornet', and 'Red Sentinel' look good in autumn as well as being functional in spring.

Other Fruit Options

In addition to the usual apples, pears, and plums, there are other fruits worth considering for your climate change garden.

Almond: More people are growing almonds. They are surprisingly easy to grow, but crops can be a little hit-and-miss, as they need a warm, dry summer. The trees like a sunny position with a well-drained soil but are drought tolerant once established. There are even varieties that survive down to about -4°F (-20°C). Try a variety such as 'Robijn' that is self-fertile, has sweet-tasting nuts, and is resistant to peach leaf curl.

Apricot: Apricot trees need an open, sunny location with well-drained soil. In the right place, a five-year-old apricot tree can potentially yield 500 fruits or more. Look for North American varieties that end in *-cot*, such as 'Flavourcot' and 'Tomcot'. The risk with apricot is the early flowering. Sally's fan-trained apricot on her garden wall is often flowering in late winter and she has had to resort to hand pollination due to lack of insects.

RIGHT: **Apricot trees may become more reliable.**

The crisp and juicy Asian pear.

Cherries require fewer chill hours than apples.

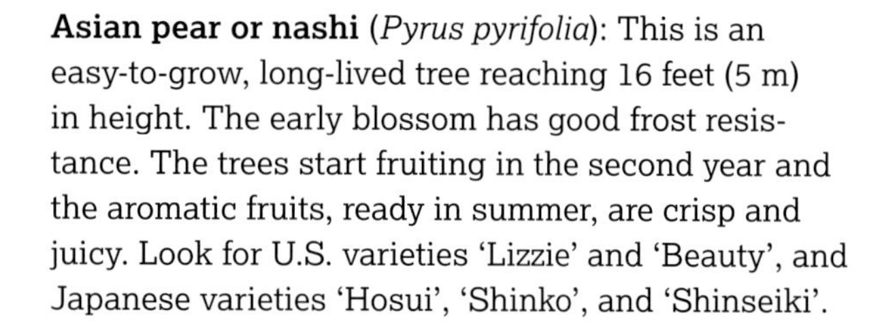

Asian pear or nashi (*Pyrus pyrifolia*): This is an easy-to-grow, long-lived tree reaching 16 feet (5 m) in height. The early blossom has good frost resistance. The trees start fruiting in the second year and the aromatic fruits, ready in summer, are crisp and juicy. Look for U.S. varieties 'Lizzie' and 'Beauty', and Japanese varieties 'Hosui', 'Shinko', and 'Shinseiki'.

Cherry: Cherry trees have a lower chill requirement than apples. Those varieties suited to Australia, California, and the Mediterranean only need 300 chill hours while others need 700+ chill hours. The downside is that they may be susceptible to cold, wet winters and frost, as the blossom is early.

Fig: This fruit needs a warm, south-facing wall as heat is needed for the fruit to ripen. It is recommended that the roots are restricted to limit the growth of the tree and stimulate the formation of fruit. This happens naturally at the base of a wall, but you could achieve the same by planting it in an open-bottomed box or in a pit in the ground bounded by paving slabs, or even by growing in a container. The plants are self-fertile. There are a number of hardy varieties, including the widely grown 'Brown Turkey', 'Brunswick', and 'Violetta'. In colder climates, fig trees will have to be protected over the winter.

Hardy kiwi (*Actinidia arguta*): These vigorous climbing plants need space and a sunny location. Hardy kiwi plants are either male or female. Only the females produce fruits, but one male vine is required for pollinating three or four female vines. The young shoots are at risk of frost damage and need protection. Small, smooth green fruits appear three to four years after planting. They do not ripen on the plant, so need to be picked and left to ripen indoors. There are also fuzzy kiwi (*Actinidia deliciosa*) varieties you can grow, but they are not as winter hardy. 'Jenny' is a good self-fertile selection.

Peach: These fruits need a warm, sheltered spot, but in mild areas you may get away with growing them as a bush. Like apricots, they need a short chill period for flowering. Flowers appear early and are at risk of frost, so they need to be protected with a row cover. The trees also need protection from spring rain, which can bring on leaf curl.

Pear: Pears have frost-hardy blossoms, but they are grafted onto quince rootstocks that need moisture, especially in spring. In a dry spring, it's recommended that pears are watered from the time the flower buds appear for six weeks to improve the yield. In a dry summer, continue to water young plants frequently. Quince C rootstock produces a smaller tree that fruits more quickly.

Plum: Plums and damsons need the most chill hours, so they are most affected by mild winters. A close relative, the gage, may be a good alternative. As well as being incredibly tasty, they are suited to drier climates. French varieties of gage include 'Oullins gage' (yellow) and the self-fertile 'Reine Claude de Bavay'.

Damsons have a high chill requirement.

Pomegranate: This plant is surprisingly hardy, with varieties such as 'Provence' being hardy to 5°F (-15°C). The self-fertile plants should be given a sunny, sheltered spot. One of the difficulties with this fruit is that it needs heat for the fruit to ripen, so long hot and dry summers followed by warm autumn temperatures of 55°F to 61°F (13°C to 16°C) are needed for harvesting late in the season.

Fruit Trees in Containers

If you have a small garden and no space for an apple tree, then what about growing a fruit tree in a container? It needs to be grafted onto dwarfing rootstock. Dwarf trees may not live as long as some of the more vigorous rootstocks, but they start producing fruit more quickly. You can also grow cherry, pear, plum, peach, nectarine, and apricot on dwarf rootstocks as well as fig and lemon.

Ideally, the pot needs to be at least 20 inches (50 cm) in diameter. Fill the pot with a good potting mix blended with grit and mulch with gravel or chipped bark. Feed every two weeks with a liquid tomato or seaweed feed and place in full sun in a sheltered position. In autumn, the less hardy apricot, peach, and lemon can be moved under cover or protected. It's important not to let the roots become pot-bound, so move the tree to a larger pot every other year or prune back the roots. The key to successful pot fruit is careful watering—not too much, but don't allow the potting mix to dry out.

Growing Olives

Olives can be grown more widely in a warmer world. Traditionally found growing high on slopes in southern Europe, olive trees can withstand a chilly winter. In fact, like many other tree and shrub species, they actually need a chill period (under 50°F [10°C]) of at least three months, with fluctuations between day and night temperatures, in order to flower and fruit, which is why container-grown olives kept in greenhouses are unlikely to flower. Equally, prolonged cold can also inhibit fruiting. Those grown outside will flower easily, but a good summer is needed to get the fruits to ripen.

Olives grow best in poor sandy-type soils in a sheltered, sunny location. They will grow in clay soils but avoid areas that get waterlogged in winter, as they don't like it cold and wet.

Most garden olives are grown in pots, in which they do well because of their small root systems and drought-hardy nature. Give them a large pot with lots of drainage holes, filled with potting soil, compost, and grit, or alternatively, create a bottomless box around the root ball with timbers or paving slabs.

Olives should be watered and fed from late winter through spring to encourage fruiting and should not be allowed to dry out in summer, as the fruits will shrivel and drop. Trees start producing fruits when they are three to five years old, fruiting on the tips of the previous year's growth, so any pruning will reduce fruiting. The crop should be thinned to three-quarters of the fruits per 12-inch (30 cm) length of branch within three weeks of flowering to ensure a crop that ripens and doesn't drop.

Olives for Eating or Oil?

You can grow olives for eating and pressing. Remember that ripe olives when eaten raw will taste quite different to commercial olives, as they are picked and steeped in water to remove the bitter taste or cured in salty water. It's very easy to make oil. Simply gather the ripe fruit, crush, and squeeze through a press. Newly pressed oil is very green initially but fades to golden yellow. You will need about 11 pounds (5 kg) of fruits to make 1 quart (1 L) of oil.

Keeping a potted olive watered in winter is also important, so don't let it dry out. Growers have found that olives can survive extremely low temperatures for several weeks as long as they are watered. Prolonged cold weather will cause the leaves to drop and the bark to split and there may be dieback in young plants. However, cold-damaged plants should regrow from dormant buds, but won't flower and fruit as well in the following months.

Which Olive to Grow?

There are more than 800 cultivars, and as you would expect, there are some that are more cold hardy or tolerant of wet soil than others. Most can tolerate 30°F (-1°C) and some are hardier still. If in doubt, grow in containers so they can be moved into a protected space when needed.

'Arbequina': A small, self-fertile Spanish tree of medium vigor, cold hardy and adaptable. It's able to cope with poor soil and is a good pollinator. The small fruits appear from four years and are good for olive oil.

'Cipressino': From Puglia, Italy, with a vigorous upright habit. It's very hardy and copes well with coastal conditions. Its black olives make a fine olive oil.

'Frantoio': A small to medium tree with large fruit from Tuscany, Italy. It's thought to be adaptable to temperate weather.

'Leccino': This popular open, semi-pendulous variety from Tuscany is easy to grow and tolerant of a wide range of conditions. It's self-sterile, so needs a pollinator and produces high-quality olives for eating and oil.

'Maurino': A Tuscan variety that will cross-pollinate 'Leccino'. The oil is delicate and soft.

'Pendolino': This Tuscan olive has a compact weeping habit, so ideal for small places. It's self-sterile and needs a pollinator. It produces black olives for eating and oil.

'Picual': (also known as 'Blanco', 'Nevadillo', and 'Picua') This Spanish olive adapts to diverse environments and responds well to regenerative pruning if damaged by snowfall. It has a good resistance to cold, bacteria, and damp soils. It is self-fertile, but also pollinated by 'Leccino'. High yielding and good for a fruity, aromatic oil.

Olives grown in large pots.

Olives may become a common sight in gardens.

It's perfectly possible to establish a small vineyard in a home garden.

Grapevines

As the climate warms up, the ability to grow grapevines in new areas increases. So, are grapes a good option and could you plant a micro vineyard?

Many people already grow grapevines in glasshouses and conservatories, but for planting outside you need to have the right spot. The ideal location is somewhere warm, sunny, and sheltered with well-drained soil, or growing up a pergola or gazebo in a sheltered, sunny spot. Like many Mediterranean plants, grapevines can tolerate the cold, but need a free-draining soil. Once they are established, the vines don't need watering and can cope with infertile soils, as their deep roots will penetrate into the subsoil, bringing up nutrients, so they don't need feeding. In fact, too much nitrogen results in more leaf and less fruit.

There are two types of grapevines: one that produces dessert grapes and the other for wine. Some dessert varieties to consider include 'Boskoop Glory', 'Muscat Hamburg', 'Strawberry', and 'Brant'. There are also some dual-purpose grape varieties, such as 'Phoenix', 'Regent', and 'Saint Theresa', that are sweet when grown in a greenhouse but outside they are better for wine.

Micro vineyards are popping up in gardens, and even community gardens, and there's likely to be even more in the future. You can plant as many as thirty vines in a 538-square-foot (50 m²) plot. Given that a well established vine can produce 3 to 4.5 pounds (1.5 to 2 kg) of grapes, enough fruit for two or three bottles, you could make upwards of sixty bottles of wine a year from your tiny plot.

TREES FOR THE FUTURE

Walk along an urban street in the height of summer and you will be very grateful for the cooling shade cast by trees. They are an important feature of many towns and cities. It has long been known that trees can help reduce the temperature by sheltering the ground from direct sunlight and cooling the air through the evaporation of water from their leaves (transpiration). They are so effective that the difference in temperature under trees compared with a nearby unshaded area on a hot summer day can be as much as 25°F (14°C). There are further benefits. Trees can be planted so that they shade buildings and reduce the demand for air conditioning in summer. Cities are already several degrees warmer than the surrounding areas, and they are going to experience even higher temperatures as a result of climate change. By increasing the green canopy of trees—the so-called urban forest—cities can help offset some of these changes.

Looking ahead, trees are an important feature of the climate change garden also, providing essential shade in summer and valuable shelter from wind, slowing water, and reducing stormwater runoff. Forestry and gardening organizations around the world are looking ahead to how trees will fare in the future, and some of their research can help inform our own decisions about which trees we should plant in our outside spaces.

LEFT: Planting trees can help mitigate the effects of heat islands in cities.

Changing Conditions

Trees are long-lived features of gardens, parks, and the wider landscape. An oak tree planted today could still be around in several hundred years' time. Therefore, the conditions that the adult tree experiences throughout its life are going to be very different than those of today. It's expected that trees will grow faster because of the longer growing seasons, the added warmth, plus the higher levels of carbon dioxide that will boost photosynthesis. But long summer droughts could restrict summer growth and cause cracks in timber, making the loss of branches more likely. Buildings near trees may be at greater risk of subsidence as trees extract more water in a dry year. Buds may open early, but this puts susceptible trees at greater risk of frost damage, especially those growing on south-facing slopes.

With an increase in frequency of storms and strong winds, it is likely that more trees will be damaged or even uprooted, especially if winters are wetter and soil waterlogged. The impact of climate change will be seen in declining tree health, difficulty in getting trees to establish, and the death of mature trees as a result of environmental stress.

A warmer spring and autumn mean that insect pests may be able to have more broods of young per year and this will lead to more damage to trees in the form of defoliation and so forth. Insect pests that appear in summer may be able to survive the winter and become active earlier the following year.

Wet winters may bring more disease, especially fungal diseases, which could take advantage of trees that are weakened by drought and quicken their death. Most at risk are tree species that are least tolerant of drought and those growing in shallow, well-drained soils. In contrast, trees growing in more northerly regions may gain from a longer growing season and warmer temperatures and put on more growth.

What Should We Be Planting?

Looking ahead, it's expected that many areas will experience warmer, drier summers and milder winters, so we need to think about the species of trees that we plant now.

On a commercial scale, foresters are already looking at the species they are planting now for harvesting in a hundred years or more, so they will be able to cope with the changing conditions. There will be changes in the makeup of the forests. But on a positive note, temperate forests are becoming more productive, and they are capturing more carbon dioxide.

For example, in the south of England, foresters are thinking about planting more beech (*Fagus sylvatica*), wild cherry (*Prunus avium*), Norway maple (*Acer platanoides*), sessile oak (*Quercus petraea*), sweet chestnut (*Castanea sativa*), Corsican pine (*Pinus nigra*), and Scot's pine (*Pinus sylvestris*), but by 2080 under the worst-case scenario, even the most resilient of species, such as beech, sessile oak, and Scot's pine may fail. Meanwhile, some common species, such as downy birch (*Betula pubescens*) and sycamore (*Acer pseudoplatanus*), may not fare so well, and Japanese larch (*Larix kaempferi*) and red cedar (*Thuja plicata*) are expected to be unsuitable for use in England by 2080.

Oaks Under Threat

A 300-year-old veteran oak, once a mere acorn, will have experienced many storms, droughts, and floods during its long, majestic life. Now it faces a new period of rapid environmental change and new pests and diseases that could easily threaten its continued survival.

It was the arrival of Dutch elm disease in the 1970s that virtually wiped out the elm across Europe, changing the landscape of the countryside. The disease, spread by elm bark beetles, has wiped out millions of elms across North America too. More recently, ash dieback (Chalara disease) has emerged as a major threat.

Sadly, the health of oak trees across Europe has been declining due to the threat of acute oak decline, which is bacterial in nature. It affects many oak species of the genus *Quercus*, including *Quercus robur* and *Q. petraea*. The disease was first identified in the United Kingdom in 2008 and since then affected trees have been found across central and southern England, as well as mainland Europe. The causes are unclear, as it seems to be a multifaceted problem, but experts agree that the general health of the tree is important for overall resilience against such threats. Soil, soil life, root health, waterlogging, drought, and the presence of pollutants in the air and water can all weaken the tree and affect its ability to withstand disease. Research shows the infected trees tend to be found in drier areas with longer and warmer growing seasons, on ground that is seasonally waterlogged or has high clay content. There is also a link to a higher levels of atmospheric pollutants, such as nitrous oxide and sulfur.

Air temperatures are much cooler under trees due to evaporative cooling.

Elsewhere, two other pathogens, known as sudden oak death and oak wilt, are having a devastating impact on oak forests across North America.

Foresters elsewhere in the world are experiencing similar problems—what species should they plant? Research suggests that between 70 and 100 percent of New England's coniferous forests may be replaced by deciduous species by 2085. In Canada, researchers are growing, testing, and sampling thousands of seedlings and studying their genomes in order to identify those that are more resistant to drought and other characteristics. This is particularly important for spruce, Canada's most important forestry species with more than 360 million seedlings planted each year (see sidebar at right).

It will become increasingly important to source seed or saplings carefully. Normally, the advice is to choose local sources as the trees will be better adapted to the conditions, but as the climate changes, it may be better to source from an area that has a climate similar to the predicted climate of the future. Despite the risk from importing pests and diseases, many foresters are recommending that material be sourced from 2 degrees latitude south of the site where they will be planted. For example, if you live in the south of England, source from France, while if you live in Pittsburgh, Pennsylvania, you may be wise to source from Virginia or even South Carolina. In fact, research finds that tree seeds usually do best when moved north, so choosing seed from a more southerly location could be a wise move.

Assisted Migration

Foresters in British Columbia, Canada, grow trees on 120-year rotation. They are having to find species that will be more suited to growing in a world that may be as much as 7°F (4°C) warmer in a hundred years. To inform their choices, government researchers have set up a trial called Assisted Migration Adaptation Trial to help forest managers understand species' climate tolerances and select seed best suited to the future climate. Seed collected from sites in British Columbia is established in test sites from California to the Yukon to see how the resulting trees adapt to changing conditions.

Moving seed within a species' geographic range or taking seed from a closely related subspecies that grows in a different climate zone does not pose that much ecological risk. However, there is far more risk when taking seed from a species with a southerly distribution and establishing it in a new area in the north, as there is huge potential to disturb the native flora and fauna, carry disease, or become invasive (see chapter 7). But maybe the need to adapt to climate change outweighs the risk?

London plane trees are a popular urban street tree.

Urban Trees

Trees in urban areas are vital. It's not just their cooling effect; they also intercept rainwater, help prevent flooding, sequester carbon dioxide, remove air pollution, and boost biodiversity as they provide habitat for many animals. They are a lifeline for our cities. But our urban areas are already experiencing change. As mentioned on page 17, the urban heat island effect means that cities are already as much as 18°F (10°C) warmer than the surrounding rural areas. This means that trees are really going to be under pressure.

Many of our best loved and iconic city trees are under threat—the linden (lime) trees of many European cities such as Berlin and Prague, plane trees in London, and palm trees in Los Angeles, for example. In many cities, the trees are mature and coming to the end of their lives, so it's a good time to consider which tree species should replace them. And as cities try to increase their canopy cover to help combat climate change, planners need to consider which trees would be best suited for the future conditions.

Studies find that the diversity of trees in urban areas is very poor. In the past, landscape architects have tended to opt for tried-and-tested species, but as conditions change those trees may no longer be suitable, while the lack of diversity leaves the urban trees open to new pests and diseases.

One of the most popular urban trees is the London plane (*Platanus* × *acerifolia*), a hybrid between oriental plane (*P. orientalis*) and the American plane (*P. occidentalis*) and a relative of the sycamore. It's a tall, hardy tree with a straight trunk and spreading branches that cast shade. It's an easy-to-grow, hardy tree; copes with most soils, compaction, and waterlogging; likes full sun or partial shade; can cope with high levels of air pollution; and it's long-lived. It is the perfect urban tree! Not surprisingly, it's found in many cities across Europe and North America, even Australia. But will this continue? As the climate warms up, it may not be the best choice. In Melbourne, for example, the London plane makes up two-thirds of the city's trees, but as they reach the end of their lives and need replacing, city planners are moving to species more typical of warmer climes, such as Moreton Bay fig (*Ficus macrophylla*) and jacaranda (*Jacaranda mimosifolia*).

Another problem facing planners is the process of getting new trees established. It takes some tree species decades to reach maturity and in that time they may be damaged by vehicles, pedestrians, and compaction, and even poisoned by salt that washes off the roads in winter. It's easy to plant a tree but far more difficult to take care of it for the next ten years or more. So we may see planners move to smaller, faster growing trees, such as the wild cherry (*Prunus avium*), that are more suited to built-up areas with little pavement space, require less water, and can deliver the benefits more quickly. Another alternative might be to opt for fruit trees that mature quickly and provide city inhabitants with food too, although no doubt some people would complain about fruit making a mess on the sidewalk!

Switch to Near Relatives

Landscape architects often choose native species, thinking they are best suited to the conditions, but as the conditions change, they may not be ideal choices. People get concerned about the use of nonnatives, arguing that nonnative species do not support wildlife, but if the native species can't survive, then there may be no option. We have to remember that the flora of a region is never constant, as new species arrive all the time and others disappear. Twenty thousand years ago during the last Ice Age, glaciers covered much of northern Europe, Canada, and parts of the United States, and as the ice retreated new species moved north to colonize the land. So, change is natural.

A good option, however, may to be look to the near relatives that grow in warmer climes. For example, the common linden (*Tilia × europaea*) is found in many cities, but it can't tolerate drought and gets stressed in a dry summer. An alternative might be the silver linden (*Tilia tomentosa*), which is more drought tolerant. This tree gets its name from the silver hairs on the underside of the leaves, it's the plant's way of deterring pests, such as aphids, so an added benefit of this species is no aphid infestations that are responsible for the honeydew that drips from linden trees in summer.

Experts such as Henrik Sjöman suggest that alternative species to trial might include the Hungarian oak (*Quercus frainetto*), an upright tree with a light canopy that is more suited to warmer cities than the English oak (*Q. robur*); the Turkish hazel (*Corylus colurna*), with a narrow, dense crown, which is more tolerant of difficult growing conditions than the hazel (*Corylus avellana*); and Japanese rowan (*Sorbus commixta*), which would be a good alternative for the rowan (*Sorbus aucuparia*), as it's a small, tough tree that's ideal for small spaces.

Another to consider is the poplar (*Populus* spp.). Poplars are fast-growing, pioneer species that are easy to grow, adaptable, and can cope with tough conditions, such as a bare, brownfield site, and once they get established their canopy will help others become established.

Jun Yang of Tsinghua University (China) looked at the impact of climate change on urban tree species in Philadelphia and his findings show that species such as eastern white pine (*Pinus strobus*), black cherry (*Prunus serotina*), hemlock (*Tsuga* spp.), and America beech (*Fagus grandifolia*) are likely to decline as the climate warms, while blackgum (*Nyssa sylvatica*), hackberry (*Celtis* spp.), honey locust (*Gleditsia triacanthos*), and red maple (*Acer rubrum*) will cope better with the new conditions. On a positive note, the warmer conditions going forward will allow for a wider choice of urban tree species across most northern cities.

The Role of the Arboretum

The role of arboreta around the world is becoming increasingly important as we try to predict which species will grow well in the future. Many different species are grown in arboreta and their experts are well placed to judge which could be grown more widely and successfully in the region.

Italian cypress may thrive in a warmer U.K.

For the United Kingdom and northern Europe, some of the species that may thrive in our parks and gardens in thirty to fifty years include:

- Golden mimosa (*Acacia baileyana*) from New South Wales, Australia
- Red alder (*Alnus rubra*) from western North America
- Paper birch (*Betula papyrifera*) from North America
- Shagbark hickory (*Carya ovata*) from the eastern United States
- Turkish hazel (*Corylus colurna*) from southeast Europe
- Italian cypress (*Cupressus sempervirens*) from the Eastern Mediterranean
- Cider gum (*Eucalyptus gunnii*) from Tasmania
- Oriental beech (*Fagus orientalis*) from eastern Europe and western Asia
- Green ash (*Fraxinus pennsylvanica*) and white ash (*Fraxinus americana*) from North America and narrow-leaved ash (*Fraxinus angustifolia*) from central and southern Europe
- Ginkgo (*Ginkgo biloba*)
- Black walnut (*Juglans nigra*) from eastern North America
- Patagonian oak or noble beech (*Nothofagus obliqua*) from South America
- Hop hornbeam (*Ostrya carpinifolia*) from southern Europe
- Paulownia (*Paulownia tomentosa*) from China
- Serbian spruce (*Picea omorika*) from the Balkans
- Wild pear (*Pyrus pyraster*) from Europe
- Tulip tree (*Liriodendron tulipifera*) from eastern North America

For North America, some of the species that may cope in more northerly latitudes of the continent include:

- Miyabe's maple (*Acer miyabei* 'Morton')
- Red maple (*Acer rubrum*)
- Ginkgo (*Ginkgo biloba*)
- Scarlet oak (*Quercus coccinea*)
- Bur oak (*Quercus macrocarpa*)
- Red oak (*Quercus rubra*)
- Bald cypress (*Taxodium distichum*)

Botanists often notice that some species that are restricted to certain soils or habitats may grow well in an arboretum in a completely different part of the world. Often a species' restricted distribution is due to other factors. For example, the fact that they can survive in an alkaline soil puts them at a competitive advantage over other species. They could grow in other soils or even other habitats, but they are not able to outcompete other species. Sometimes there are surprises. For example, a specimen may survive a drought or an extremely hard winter against the odds, so that is the tree from which to save seed or take cuttings.

Cities around the world are working toward greater canopy cover to bring the many benefits of trees. Ultimately, to hedge our bets against the uncertainties of the future, it's best to plant lots of different kinds of trees, as we cannot be certain which species will thrive and which will not.

Planting New Trees

Choosing the best time to plant your tree is your first decision. Bare-rooted trees are only sold in autumn and winter and have to be planted immediately or heeled into the ground until conditions are right. Container-grown trees can be planted all year, but they will need a lot of care if planted in late spring and summer. Ideally, you want to plant your new tree when the soil is moist, but not waterlogged, and you definitely don't want to plant while the ground is frozen. If you plant too early in the winter, you risk your tree sitting in cold, waterlogged ground for the rest of the winter.

In recent years, Sally has struggled to find the right slot to plant her new fruit trees on her heavy soils. Torrential rain in late fall and early winter has left her soil waterlogged for much of winter, so she has ended up planting in late winter, which didn't give the trees long to get established before the weather started to warm up.

Drought is always a concern after planting a new tree, and it's increasingly likely that your new trees will have to cope with a hot summer, so a good start means they can establish a large root system as soon as possible. You will also have to make sure that drought or waterlogging doesn't restrict early growth.

As mentioned already, you need to make sure that the soil is well drained and has plenty of organic matter. The soil should be watered before you plant, as you don't want the plants struggling to take up water. It's also beneficial to soak bare-rooted trees for thirty minutes before planting and to water a container-grown plant. The hole you dig needs to be no wider than twice the diameter of the root ball, but no deeper than the root ball itself. Make the hole square too, as that helps the roots grow into the surrounding soil. Also, it is important to make sure the soil around the hole is not compacted; if it is, the water will sit around the roots and not drain away, so fork the soil around the hole, but don't till it.

Position the tree in the hole. If it's a container plant, scrape away loose soil on the top and tease out the roots. Make sure the tree is at the correct level in the ground, never deeper than the original level, so look for the mark on the stem indicating where it started to grow above ground. Sprinkle the roots with mycorrhizal powder and backfill the hole with soil, gently treading the ground to get rid of any air holes and pockets where water could collect. At this point, you don't want any organic matter in the hole, as it has a tendency to shrink and allow water to collect inside. Instead, spread a layer of mulch around, but not against the tree's trunk, to keep in moisture and suppress weeds. You may also need to apply rabbit or deer protection.

Mycorrhizal Root Treatment

A lot has been written about the benefits of applying a mycorrhizal fungal powder to roots before planting. Some gardeners swear by it and apply it to the roots of all their transplants and even use it on seeds, while others consider it a waste of time. The USDA Forest Service has found inoculating seedlings prior to planting has increased early growth.

The theory behind the treatment is sound: you want your tree roots to associate with mycorrhizal fungi as soon as possible. The fungi are fast growing and will soon produce a network of hyphae through the soil to take up water and nutrients and give the tree a good start. It's thought to be particularly important for bare-rooted stock, as some of their roots will have been damaged when they were levered out of the ground. Typically, these trees suffer from transplant shock and the plant is unable to take up water and nutrients from the soil until its roots have started to grow. A treatment of mycorrhizal fungi can help container-grown trees and shrubs too. These plants can be slow to get started, often because they don't have a large enough root system to support the shoots. A large container-grown plant may be rooted in a small pot, so the roots take time to settle and start growing into the surrounding soil and this is achieved more quickly with mycorrhizal fungi.

There is no doubt that the presence of these mutualistic fungi is beneficial to most plants and the soil. The fungal hyphae extend through the soil, binding the soil particles together and creating a better soil structure and improved water-holding capacity, which provides better drought resilience. However, you don't have to buy commercial powders. Any soil that has had plants growing in it should have fungal spores present, as will a compost rich in woody materials. And if you don't have any compost, you could collect some leaf mold from under trees and hedgerows and use that as an inoculum for the new plant.

Commercially, there are mycorrhizal fungal products with appropriately sourced mycorrhizal fungi, including powders that you sprinkle over roots or make up into a drench or gel. The idea behind these products is to treat the roots of new trees before they go into the ground, so the roots are pre-inoculated with the right fungi.

Sugar and Biochar

Another treatment that may be effective in helping your tree to establish is a sprinkling of sugar around the roots before you backfill the hole—yes, sugar! This has been found to stimulate the growth of mycorrhizal fungi and may even fuel the growth of new roots. And another option is to add biochar to the planting hole (see chapter 4). The recommended dosage is 5 to 10 percent biochar by volume in the planting hole before you backfill. Unlike organic matter, biochar doesn't degrade, so there is no loss of soil volume.

Staking Your Trees

It is always recommended to stake your new tree at the same time as planting unless it's a small sapling. This is because a tree can take several years to establish a decent root network and to anchor itself firmly in the ground, so a stake prevents the plant from moving in the wind and damaging new roots. Once you are certain the plant is well established, remove the stake.

RIGHT: **Newly planted tree with stake, tie, and mulch.**

Aftercare

There are a few things you can do to ensure the survival of your tree. Make sure the area around the trunk is well mulched to retain moisture and keep down weeds. It's important to water your new plants through their first summer, regardless of the weather. Often the surface layer can be moist, but deep down the soil around the roots may be dry. Some people recommend irrigation tubes to get the water straight to the roots, but it's better to apply it to the surface of the soil with a watering can fitted with a rose to mimic rainfall. That way the water percolates slowly through the soil. Do take care though, as you don't want to add too much—the young roots won't extend out that far in search of water. You can check you are using the right amount as the water should drain away within ten minutes.

A Low-Carbon Way of Cooling Your Home

Trees can be effective in reducing the heat load on your home in summer. A shaded wall can be several degrees cooler than an unshaded wall and is far more effective than using curtains to shade a room. Trees can be planted on the east, south, and west of a house: east planting will shade the house in the morning, although the temperature reduction is not so great. South planting means shade in the late morning to early afternoon, but trees need to be planted near the house or they have to be taller species so they cast a long enough shadow to shade your home. You need to avoid this location if you have solar panels on your roof or a passive solar system in the house. West planting is ideal, as the tree casts a shadow during the afternoon when temperatures are peaking and so will have the greatest effect on reducing temperatures inside the house. You can plant smaller trees or large shrubs on this side, as the sun is lower in the sky and the trees cast longer shadows.

A deciduous tree with a heavy canopy is best. It casts a dense shade in summer and because they lose their leaves, the sun can still penetrate the branches in winter to warm up the house. Many landscape architects recommend growing slower-maturing species. Although they take longer to reach full height, they tend to have deeper roots and stronger branches, so they will be less prone to drought and wind or snow damage.

If you don't want to plant a tree or need a quick fix, using trees in pots and other large containers is another option. A container-grown tree of around 6½ feet (2 m) in height planted a few feet (m) from the house will start to shade the ground-floor windows in its first year.

Natural shading from trees reduces the heat load on a house in summer.

Shady Areas

As temperatures increase, we will be looking to create more shade in our growing spaces, not just so it's more comfortable for us, but also to shade other plants and the soil. Trees can cast shadows over paths, driveways and patios to provide welcome shade on the hottest of days. Shady areas increase the diversity of conditions in your garden and allow you to grow a wider variety of plants. Aim to create a dappled rather than dense shade so that enough light reaches the ground to enable you to have an understory, while reducing heat stress and cooling the soil. Trees with an open canopy let the light penetrate. The trees will intercept some of the rain, so it will be necessary to mulch and water in summer.

What About Autumn Color?

One of the delights of autumn is the change in leaf color. For many gardeners, autumn color is a factor when choosing trees and shrubs for the garden, while in parts of North America, it's a major tourist attraction. So how will climate change affect autumn color? In New England, leaves start to change color from mid-September and reach a peak in mid-October. It is such big business that there are foliage maps and hotlines to call to help you see the best colors available.

Each year is different and estimates of peak color are based on the temperatures and rainfall of summer. A cooler summer will delay leaf changes, while a hot, dry summer will bring it forward. A wetter-than-usual summer delays the color, but it will be brighter. However, too much rain, with floods and waterlogged soils, stresses the trees and the leaves change color early.

Now climate change is pushing back the date of peak color. Studies show that the start of autumn has been delayed by a few days in Pennsylvania to fourteen days in New England. Not only will color change be later, but experts predict that an early spring combined with a drier, hotter summer and an extended growing season into autumn will result in more muted colors overall.

Autumn colors are formed from pigments in the leaf. There are three main pigments: chlorophyll (green), carotenes (yellow), and anthocyanins (red), with chlorophyll masking the presence of the other pigments so leaves appear green. In autumn, the shorter days and cooler nights initiate the breakdown of chlorophyll, so the green color disappears and the

carotenes become visible, turning the leaf yellow and orange. In preparation for leaf fall, a corky layer forms across the leaf stalk and this stops sugar from being moved out of the leaf. Instead, it gets converted into anthocyanins and the leaf takes on a red to pink appearance.

If autumn is warm, leaves continue to photosynthesize even though the sun is low in the sky and the days are shorter, but they produce less sugar. Not only is the autumn color delayed until October or November, but less sugar means less anthocyanin and more muted colors. A wet, cold summer also leads to poorer color because the trees have produced less sugar. If spring is early and the summer dry, the leaves will have been very stressed and trees are likely to drop their leaves early before the color changes are complete.

THE FLOWER GARDEN

Our gardens are going to look very different in ten or twenty years' time. That sounds a long way off, but for a garden it's not, especially when you consider it can take several years to get borders looking good. And with the extreme weather, getting plants established may be more difficult, so we'll have to get used to having more failures alongside our successes. It is definitely going to be a period of trial and error to see what works and what doesn't and adapting to the changes that will keep on coming as the world gets warmer and warmer. You will find increasingly that plants that we were told would never survive in your region will do very well, while some of our favorites may find conditions tough.

We use a diverse range of plants in our gardens—trees, shrubs, herbaceous perennials, biennials, annuals, roses, climbers, grasses, and bulbs. It's this diversity that will help us as we move into more uncertain gardening times. We may not be able to grow all our favorites in the decades ahead, but we may be able to capture the "look" with a different selection of plants. A popular mantra is "right plant, right place," but now we need to rethink the right place bit in its entirety!

LEFT: A cottage garden with a mix of flower types.

The Next Decade May Be Tricky

When you are buying new plants for the garden, replanting a bed, or establishing a new feature, try to choose species that will thrive into the future and group them according to their needs. Where possible, choose tough, drought-resistant perennials, trees, and shrubs that will be able to cope with the drier summer conditions. The more-demanding plants will need extra care and water. Aim to create a resilient garden by knowing your plants and opt for some wild areas that encourage wildlife (see chapter 6). Try to manage weeds and pests naturally rather than resorting to chemicals that will destroy the natural balance in your garden, especially in your soil. Also cut back on inputs such as fertilizers that not only encourage lush growth but also create a much needier outside space.

Tree ferns are cold hardy, but they don't like heat either.

Do I Choose Frost-Hardy Plants?

We know that the chances of cold winters where temperatures fall to 14°F (-10°C) or lower are going to become increasingly less common, as are late spring frosts, but there is going to be the odd one that will catch you off guard. It may not be the occasional low temperatures that kill off your less-hardy plants, but a prolonged freeze or, more likely, a wet winter when a cold snap freezes already wet ground. So, for the next few years, it will be best to play it safe by buying hardier varieties. You can also wrap up your valuable less-hardy specimens, such as tree ferns and large palms, or move them under cover just in case.

A traditional border against a wall with a mix of species, some sun lovers, some not.

The Flower Border

The flower border has its origins in the English cottage garden, with its mix of flowers for the vase and vegetables for the pot. It's a loose style of more natural planting using plants that vary in shape, color, and texture. They are at their best in summer. The most common element of the border is the herbaceous perennial—a hardy plant that dies down in the autumn, surviving underground and reappearing in spring, when it produces a rapid growth of shoots and then flowers. Perennials are long-lived features of the border, and as their performance wanes they can be rejuvenated by being dug up, split, and replanted.

Flower borders are likely to suffer as many of our favorite plants have lush growth and require a plentiful supply of water, making many of them distinctly unsuited to the warmer conditions of the future. Some garden regulars can cope with a summer drought—aster, phlox, and lupins, for example—but others, like bearded iris (*Iris × germanica*), love the heat but hate cold, wet winter weather. However, a warmer spring may mean there is less risk of losing delphiniums to late frost.

One problem with our flower gardens is the way the planting is mixed. We gardeners often tend to plant for color or appearance and muddle up the water-hungry plants with the drought-tolerant ones. We won't be able to get away with that in the future. Our borders will have to be planted with water requirements firmly at the front of our mind.

Don't worry, it's still going to be possible to have an eye-catching herbaceous border, but the composition of species will be different. You may lose your favorites, but there will still be a huge range of herbaceous plants available to use, and most are quick and easy to establish, so unlike trees, you can redesign your herbaceous bed and see the effect in a few years.

The drought tolerant sea holly (*Eryngium* sp.) has striking blue flowers that are surrounded by spiny blue bracts.

Now might be the time to think about incorporating pelargoniums into the border, rather than growing them in pots.

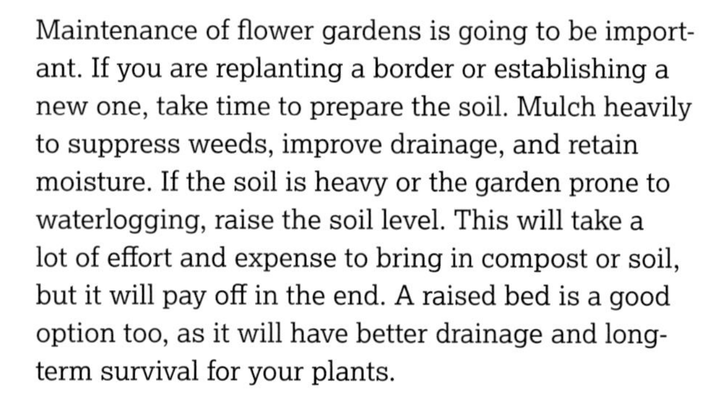

Maintenance of flower gardens is going to be important. If you are replanting a border or establishing a new one, take time to prepare the soil. Mulch heavily to suppress weeds, improve drainage, and retain moisture. If the soil is heavy or the garden prone to waterlogging, raise the soil level. This will take a lot of effort and expense to bring in compost or soil, but it will pay off in the end. A raised bed is a good option too, as it will have better drainage and long-term survival for your plants.

Often flower gardens are left to their own devices when it comes to watering, but you will have to pay more attention in a hot summer, especially if you have newly established plants. And as mentioned before, drench the borders so you only have to water every few weeks, to encourage deep rooting. You may need to pay more attention to staking due to the greater likelihood of strong winds and heavy summer downpours. Winter storms and heavy rain may result in nutrients being washed out of the soil, so an annual mulching will help retain moisture and supply nutrients.

The backbone of the climate change border will be drought-tolerant perennials. Look for the easy-to-grow, low-maintenance species that flower well and, most importantly, are resilient. A colorful, long-lasting border could comprise achillea, eryngium, helenium, Jerusalem sage (*Stachys byzantina*), lamb's-ear (*Stachys byzantina*), mullein (*Verbascum* spp.), red hot poker (*Kniphofia* spp.), *Verbena*, and *Salvia rosmarinus*.

Also, don't forget that tender perennials will come into their own in the future. Compared with more traditional perennials, such as delphiniums and lupins, these warmer climate perennials flower for many months, starting early and continuing into autumn. Already gardeners are finding that these tender perennials can overwinter, and with winters getting milder, overwintering will become the norm. You just have to guard against cold, wet soil.

Annuals can be planted in borders too, especially when the border is newly planted and there are gaps. Gazania and osteospermum cope well with heat and wind, while the fast-growing antirrhinum, petunia, calendula, and salpiglossis will all provide color for many weeks.

A Gravel Garden

If you find the maintenance of a needy lawn too much hassle, why not dig it up and replace it with a gravel garden? Gravel helps heavy rain percolate down to the water table, so the ground is less likely to get waterlogged and, importantly, plants that like it dry are not sitting with wet roots in winter. In summer, the gravel acts as a mulch to retain moisture.

To be successful, a gravel garden needs to be prepared properly because it's not going to be watered or fed beyond establishment. Many will say simply to clear away the grass to bare soil, add a weed barrier, and lay down a thick mulch of gravel, perhaps with a few rocks as a feature. However, time spent improving the soil will be invaluable, especially if the soil is compacted. Incorporate plenty of organic matter. Don't forget that there are is wide choice of gravels available, varying in color and size and some that are self-binding too. To plant, scrape away a section of gravel, dig out a hole, remove the soil, place a root ball in, and push back the gravel to leave a clean surface.

There are many drought-resistant plants suited to the dry conditions of a gravel garden. Gravel planting needs to be bright because paler, pastel colors can look a bit insipid against gravel. Some of the plants suited to this type of garden include euphorbia, lavender, nepeta, and stachys, plus grasses to give texture and movement to the planting. and if you want to create an exotic style, you don't have to look much further than agave, dasylirion, and yucca.

This drought-tolerant border in the Roads Water Smart Garden in Colorado is only watered once or twice a year if needed.

One inspirational gravel garden is the famous Beth Chatto garden in Essex, England. The garden was established in the 1990s on the site of an old carpark and despite being located in one of the driest parts of the United Kingdom, it is never watered. This is possible because it is planted with drought-tolerant species. Beth Chatto's approach to planting was to make pictures with form, structure, and color. She tended to plant in triangles, varying the height of the plants and incorporating plenty of texture by contrasting spiky with soft. She used accent colors and architectural plants to draw the eye across the planting.

Deep Gravel Planting

Gravel can be used as an effective mulch to retain moisture and suppress weeds. But there is a movement toward using deep gravel mulches, especially those made from recycled aggregates.

The naturalistic style of planting was made popular by Noel Kingsbury and Piet Oudolf (see page 193). They created low-maintenance perennial plantings by laying a gravel mulch over the soil or even the subsoil. More recently, Nigel Dunnett and James Hitchmough from the University of Sheffield in England, Peter Korn in Sweden, and others have taken this further using deep mulches of various mineral substrates.

A deep gravel mulch, often up to 20 inches (50 cm) deep, creates a sterile, low-fertility surface layer that is free from weeds. It is usually created using pea gravel, but a more sustainable alternative is recycled substrate making use of building waste, such as concrete, brick, ceramics, asphalt, and even mining waste. In some schemes, the topsoil is removed to reduce fertility and create conditions in which drought-tolerant perennial plants can thrive as well as removing a source of weed seeds.

When creating a sustainable urban drainage system with a naturalistic planting in Sheffield, Nigel Dunnett covered the soil with an 8-inch (20 cm) layer of recycled substrate mixed with some composted green waste and sandy loam soil. This layer not only enabled the perennial plants to get established by reducing competition but also helped rainwater infiltrate the ground and lower the flood risk. It also retains moisture in summer and reduces the need to weed, an important feature for urban public spaces.

The Half-Hardy Annuals

Hardy annuals typically peak in mid-summer, so they are more likely to suffer from lack of water in a hot year and go to seed, rather like some vegetable crops. Instead, in the future, it may be wise to look to the half-hardy annuals, including cleome, cobaea, cosmos, nicotiana, and tagetes. With warmer springs and reduced risk of late frost, they can be sown earlier. They could even be sown in autumn, like a hardy annual, as winters become milder. There may be opportunities to have an extra bedding season too, running from late summer through autumn into early winter, providing even more color in the garden.

LEFT: **A newly planted gravel garden in the style of Beth Chatto.**

Salvias

Salvias are generally drought-resistant and long-flowering and many have scented foliage, plus they are great for attracting pollinators. They are the perfect option for a hot, dry border with free-draining soil. There are annual, biennial, and perennial salvias. Planted in the right spot, the shrubby and hardy herbaceous salvias can overwinter. The half-hardy species need protection from frost, so are either grown in pots or lifted and moved to a greenhouse, such as eyelash sage (*Salvia blepharophylla*), Guatemalan leaf sage (*Salvia cacaliifolia*), pineapple sage (*Salvia rutilans*), and bog sage (*Salvia uliginosa*), while gentian sage (*Salvia patens*) has tubers and can be mulched or lifted like a dahlia. Some of the popular annual salvias, including mealy cup sage (*Salvia farinacea*) and scarlet sage (*Salvia splendens*) are actually short-lived tender perennials and will survive winter if given protection.

Prairie Planting

Rather than a traditional flower border, a drought-tolerant alternative is the free-form prairie-style planting, with drifts of grasses and perennials, all differing in shape, color, and flowering times. It provides interest year-round with lots of movement, layers, and viewpoints.

The best location is one that is sunny all day, with soil that's rich in organic matter and well drained. Build up the soil to ensure good drainage and start with as few weeds as possible as these beds are notoriously difficult to weed once they are established. One way is to cover the bed with plastic for a year before planting to kill the weeds.

The prairie-style planting works on all sizes of bed and can be planted with established container-grown plants or sown from seed, which creates a truly natural look, albeit more difficult to achieve. It's important to keep the planting well watered for the first year; these plants may be drought tolerant, but you want good establishment and for them to develop an extensive root system. Any gaps can be planted up with annuals. Ongoing maintenance involves mulching to both retain moisture and suppress weeds. Don't tidy the beds in autumn, but leave the old stalks to provide interest, frost protection, seed for birds, and to help with drainage. They can be cut back in spring, once the weather is kinder.

Naturalistic Gardens

Piet Oudolf is a celebrated Dutch landscape garden designer who uses bold drifts of herbaceous perennials mixed with grasses, inspired by the way they grow in the wild. He was behind the planting on the High Line (see page 90) and Battery Park in New York as well as many other gardens around the world. He is particularly interested in landscaping for public spaces that should be beautiful, yet functional and attract a wealth of wildlife. This perennial meadow in Somerset, United Kingdom, comprises a mix of grasses and perennial herbaceous plants that create a moving backdrop and among them is a selection of focus plants that add color and texture. Color in spring comes from bulbs, such as allium and camassia, but the perennial planting is at its best in late summer and early autumn. The soil has been mounded up and is gritty to improve drainage.

ABOVE: **Drifts of helenium and anemone in early autumn.**

The Dry Garden at Hyde Hall

The RHS Garden at Hyde Hall is in Essex, one of the driest parts of England and, believe it or not, the annual rainfall here is 24 inches (61 cm), less than that of Rome. Construction of the dry garden started in 2001 to show gardeners that it's possible to have a beautiful garden that doesn't need watering. The garden lies on a south-facing slope and was created with mounds of hardcore and subsoil covered with a gritty topsoil and finished with boulders and a thick mulch of gravel. Its dry environment is very reminiscent of a Mediterranean hillside. Being on a hillside, the plants are exposed to cold winds from the east and there's little shade from the sun; it has not been watered since its establishment. There are more than 400 different species, all perfectly at home in the dry environment, including agave, brachyglottis, giant viper's bugloss, euphorbias, perovskia, santolina, verbena, and grasses, such as pampas grass, calamagrostis, and stipa. There are lots of self-seeded annuals too, including eryngiums, Californian poppies, and nigella.

Crocus look lovely planted in a lawn, but will they survive a changing climate?

Can I Grow Bulbs?

Our gardens can look simply wonderful in spring with swaths of bulbs, from the first snowdrops to daffodils and tulips. But how will bulbs cope in a warmer world? Simply put, there are winners and losers.

Bulbs are storage organs. They enable plants to survive over winter and reappear each year. These perennial plants build up food stores in their bulb during the summer months and then lie dormant through winter until the right conditions return. Some species are found in woodlands and they emerge early so they can complete their life cycle before the tree leaves open and reduce the light reaching the ground. In the dry climates of the Mediterranean and South Africa, bulbs appear in spring or autumn to avoid the heat and drought of summer. A spring-flowering bulb doesn't respond to day length, but to the temperature, the size of the bulb, and its food reserves. In bulbs like daffodils and tulips, development starts after flowering is complete in spring and early summer. In contrast, in summer bulbs, such as lilies and gladioli, the flower forms after the emergence of its shoot in spring and early summer.

Daffodils don't like waterlogged soil.

Daffodils

Winter temperatures affect the timing of daffodils. Mild winter weather leads to an early flowering, whereas colder than average weather can push the appearance of the flowers back.

Daffodils are native to Spain and Portugal, so they can cope with a mild winter. They grow in most soils and need plenty of organic matter to supply nutrients, but they don't like being waterlogged. If our winters get wetter and you have clay soils, daffodils will have to be planted where soil is well drained or in raised beds. They shouldn't be planted too deep, as they have to expend more energy to reach the surface before their leaves can emerge and start to photosynthesize and replenish food stores.

Tulips

Tulips are a huge tourist attraction in some parts of the world and, if they flower too early, tourists will miss the flowers. The Tulip Time Festival in Holland, Michigan, is a popular festival. In 1920, the festival started in mid-May, and this was the same until the mid-1970s, but then the trend to early flowering got underway. Now, the festival opens at the beginning of May, two weeks earlier than in the 1970s. If the weather is warm in February and March, rather than cold or wet, the flowering period of the tulips is much reduced, so there is plenty to concern the organizers.

Tulips, like crocus and hyacinth, need a period of cold to stimulate flowering. This is called vernalization. Tulips grow in mountainous areas with a temperate climate, such as Central Asia. Because of the cold winters, the tulip's development is triggered by temperature and, if the winter is mild and wet, tulips simply do not perform well.

Temperature is key in the life cycle of the tulip. It needs warmth to start developing the flower and a period of cold to break dormancy and trigger the growth of the stem and flower in spring. If the temperatures are higher than average in spring, the quality of the flowers the following year can be adversely affected. The bulbs will be at a more advanced stage when they are harvested and put into storage and this leads to the dehydration of the flower bud.

As temperatures increase and winters become milder, it will become more difficult to grow crocus, tulips, and hyacinth in some regions because of the lack of chill factor, but they could be grown in a climate change garden if they are given six to eight weeks of chill to trigger the flowering cycle. To overcome this hurdle, it may be that we have to pop our tulips in a fridge for a couple of months and plant them out in winter or start buying prechilled bulbs.

What can you do if you want to grow tulips? One option is to plant them in shady places so they are less affected by unseasonably warm weather. Once the tulips have experienced some cold weather, or when the ground temperature has fallen, mulch heavily to maintain the cool conditions. Another way forward may be to plant a range of bulbs with differing rates of maturation so there are better chances of some flowering. There are lots of different species and varieties of tulips and some will be better suited to a warmer garden than others. Among the best suited are the Darwin hybrids, reliable tulips that are less affected by changeable weather, so not surprisingly they are popular around the world. They produce a single, large flower on a long, sturdy stem. These mid- to late-spring tulips flower for up to five years and are good for naturalizing (see page 198). Another option is to choose late-flowering varieties, such as 'Queen of the Night', 'Menton', 'Dreamland', and 'Dordogne', but their flowering period may be shortened if early spring is dry and warm.

Naturalized Bulbs

The modern bulb has been bred to deliver a great display in the first and second year, and then be replaced, while tulips grown for cut flowers are kept for just one year and then discarded. However, there are naturalized bulbs—the daffodils, crocus, and snowdrops that are true perennials and have been selected for their ability to reappear year after year. Tulips can be naturalized too, but you have to choose varieties that have been bred to naturalize or use species of tulips, such as *Tulipa clusiana* and *Tulipa tarda*, that will last many years and multiply. Being perennials, naturalized bulbs have a chance to adapt to the changing climate.

Bulbs for a Warmer Climate

Rather than attempt to grow daffodils and tulips in the future, it may be better to switch to bulbs that are adapted to a warmer climate with dry summers and mostly mild winters with lowest temperatures around 23°F (-5°C). The options include:

Allium: These bulbs flower from spring to summer. They are planted in autumn, deep in a fertile soil to give a good flowering for many years. They like full sun, but don't like cold, wet soils or exposed conditions. Some can naturalize in grassland, such as *Allium hollandicum*, while others, such as *Nectaroscordum siculum*, can be found in light woodland where the soil stays moist, but not waterlogged, and the bulbs are sheltered from the worst of late winter and early spring weather. They are a good option to plant in the herbaceous border to give early interest.

Narcissus: They originated in the Mediterranean and can cope with summer drought, as the bulb is dormant in the ground. However, they are more susceptible to wet winters with waterlogged, cold soil, as this will delay flowering.

Summer snowflake (*Leucojum aestivum*): This produces spikes of bell flowers in mid-spring and is dormant in summer. It is planted in autumn in a moist, fertile soil, well drained in sunny to light shade. It can cope with boggy areas, so could be planted around ponds and in bog gardens. It naturalizes easily and looks great under trees.

LEFT: **The species *Tulipa tarda* flowers in early spring and is ideal for naturalizing.**

Wood anemone (*Anemone nemorosa*): This hardy bulb has long-lasting daisy-like blue, white, or pink flowers. It flowers early, has a long season, and naturalizes well. It likes partial shade with fertile, well-drained soil, but will cope with full sun. It's ideal under deciduous trees and shrubs.

Other options are plants with tubers and rhizomes rather than bulbs. They include:

African lily (*Agapanthus* spp.): This rhizomatous perennial is a native to South Africa. It is planted in autumn or early spring to flower from summer to autumn. It's drought and salt-tolerant, so ideal for seaside gardens. It needs a fertile, well-drained soil and can be protected from cold weather by a thick mulch in winter, but the less hardy types in containers need to be moved under cover.

Bearded or German iris (*Iris × germanica*): This is a rhizomatous perennial from the Eastern Mediterranean and likes full sun. It should be planted in a south-facing border with a well-drained soil, but it doesn't like waterlogged soil.

Crinum or swamp lily (*Crinum asiaticum*): This produces beautiful funnel-shaped flowers held above the leaves. It needs full sun and a well-drained soil that doesn't get waterlogged in winter. It may need protection against frost in northern areas. It's ideal for sunny beds along a wall, with the bulb planted so the neck just emerges from the soil.

Nerine: This easy-to-grow late-flowering bulb from South Africa needs a sunny location with well-drained, gritty soil. It's intolerant of waterlogging and wet soils in winter, so plant bulbs with their neck just above the surface of the soil. Flower stalks appear in autumn. An ideal position is at the bottom of a sunny wall. These hardy plants benefit from some water when growing, but can be dry when dormant.

Pineapple lily (*Eucomis* spp.): These eye-catching plants with a spike of star-shaped flowers are usually grown in containers, but in a warmer garden they could be planted in a sunny border. Pick an open sunny spot with fertile, well-drained soil. They can tolerate drought, but this will adversely affect the flowers, so don't let them dry out. In winter, mulch with bark or straw or lift the bulbs and store in a greenhouse

RIGHT: **Pots of pineapple lilies (*Eucomis bicolor*).**

Succulents

Succulents are adapted to living in desert and arid habitats where there is little water and thin soils. They store water in their fleshy leaves, stems, and roots, while the waxy layer covering their leaves and stems reduces any water loss, helping them conserve water. They tick all the boxes—easy to grow, drought resilient, evergreen, and low maintenance, so expect to see more of them in gardens in the future. Their drought resistance makes them the ideal choice for a green roof (see page 93) or vertical garden.

Succulents need a sunny site and can cope with low temperatures. Deserts can be incredibly cold by night but, like many of the heat-loving plants mentioned in this chapter, they can't cope with waterlogging or wet soil in winter. If they get too wet, they will be killed by frost and snow. The right conditions can be achieved by planting them in pots, containers, or gravel beds. They grow well at the foot of sunny walls, as the wall radiates heat and helps shield them from rain.

There are places where the microclimate allows a subtropical flora to survive in temperate zones, such as the fabulous gardens of St. Michael's Mount on the Cornish Coast in the United Kingdom, where succulents grow profusely on the cliffs. Here the mild climate and exposed, rocky site is perfect for growing a subtropical flora. Frosts are rare, as the rocks act as huge storage heaters, absorbing heat during the day and releasing it at night. In summer, temperatures exceed 104°F (40°C). The cliffs are planted with an array of succulents, including agave, aeonium, aloe, and sedums, plus other exotics, including ginger (*Hedychium* spp.) and *Leucadendron* spp.

The eye-catching Aeonium or tree houseleek

A Succulent Hanging Basket

Hanging baskets are usually very demanding of water and in a hot summer there is a need to water them daily. A new twist on the hanging basket is to use succulents. Not only will the hanging basket be low maintenance, but it will also provide year-round interest. All you need to do is water occasionally in drier weather, don't water from late autumn to spring, and give them a feeding once in spring.

A subtropical planting on St. Michael's Mount off the coast of Cornwall, England.

Succulent	Characteristics
Aloe	Various types, including *Aloe vera* and *Aloe variegata*, are easy to grow and feature long fleshy leaves and toothed margins. Spikes of tubular yellow-orange flowers. Needs a sunny location and to be moved to a frost-free location in winter, so ideal for containers. Good for coastal and urban gardens.
Houseleeks (*Sempervivum* spp.)	These rosette-forming succulents are very easy, tough, and slow growing. Their summer flowers are good for pollinators. They are the extreme survivors, coping with high temperatures and drought, as well as being hardy. They can be planted in pots, bricks, walls, and gravel gardens because they can grow in very little soil.
Mexican snowball (*Echeveria elegans*)	These rounded, rosette-forming plants have pointed leaves, upright stems, and pink-coral colored flowers in late winter and spring. Plant in full sun in a southerly, sheltered aspect. Not hardy.
Stonecrop (*Sedum* spp.)	There are various kinds of sedums, including the ground cover types, such as *Sedum acre*, that are used on green roofs, and the taller *Sedum spectabile*, *Sedum telephium*, and the hybrid *Sedum* 'Matrona' (aka *Hylotelephium* 'Matrona'). The taller clump-forming succulents have fleshy leaves and stems, but they do have a tendency to collapse if they are given too much water. They need harsh conditions to create a sturdy growth. Plant in a sunny spot in well-drained soil. They are drought tolerant and work well with grasses, agapanthus, eryngium, and salvia. They don't like waterlogged or wet soils.
Tree houseleek (*Aeonium* spp.)	This striking, upright succulent has rosettes of leaves in a range of colors from green to bronze and black. *Aeonium* 'Zwartkop' is a particularly eye-catching plant with deep black-purple leaves and small yellow flowers. Aeoniums are great for courtyard gardens and frost-free locations. They need a sunny, sheltered position in summer, such as near a south-facing wall, but they are not frost hardy or tolerant of wet feet. In winter, move them to a cool greenhouse or porch. They drop their older leaves in winter, so if they start to look leggy, the top of the shoot can be cut off and rooted in gritty compost and the rest of the plant repotted and allowed to reshoot.

Can I Have a Lawn?

Many say that a garden is not a garden without a lawn, but it's going to be the first to suffer in the climate change garden. The more highly managed and manicured the lawn, the quicker it's going to suffer. We are going to have to get used to our lawns turning yellow and then brown in a dry summer; despite the horrible appearance, grass is a truly resilient plant and will recover within weeks. Just remember "brown is the new green!"

Grasses are the dominant plants of the savannah and prairies, supporting millions of grazing animals. They are tough plants with deep roots that extend many feet into the ground to reach water. The roots stabilize the soil and help build up a humus-rich earth that supports high productivity. The prairies experience a continental climate—hot dry summers, cold winters—so grasses and other plants can cope with months of hot, dry weather with the occasional storm, something to remember when we think about the future of our garden lawn.

Most lawns are comprised of a mix of grasses. Fine-leaved grasses are used to create the pristine, manicured lawn that requires mowing, feeding, weeding, aerating, and watering as soon as it's dry. This type of lawn will be the first to suffer in a drought or from flooding, while a lawn with a selection of broad-leaved grasses is more tolerant of trampling. Generally, lawns can look after themselves as grasses spring back to life once the rain returns. The lawn may not look attractive, but it really doesn't need watering.

Don't worry if your lawn turns brown.

The higher temperatures will favor different grasses. The finer-leaved grasses are the least tolerant of drought, while grasses with tough fibrous leaves are better suited. Some of the weeds will be winners too. Looking far ahead, it is likely that by 2050, we will need to be growing those grasses typical of southern Europe or Florida!

Climate Change and Lawn Mowers

While on the topic of lawns, think about your climate change credentials. How much fuel do you use to cut the lawn? The best climate-friendly option is not to cut your lawn and have a flower meadow instead! Next best is the push lawn mower. These are much improved design-wise, and they have the added benefit of affording a good workout too, with all the pushing involved.

After that you might consider the mulching lawn mower. This mower chops up the grass clippings and drops them on the lawn to decompose and recycle the nutrients. They are efficient energy-wise as they reduce the time taken to cut the lawn by as much as half, so there is less fuel used per mow. There are some cordless mowers powered by lithium-ion batteries, so although they require electricity to recharge them, it's more efficient than using a gasoline engine, especially if the electricity is supplied by renewable energy sources.

LEFT: **Consider lawn alternatives, such as groundcovers. They don't need to be mowed and many are drought-tolerant.**

Lawn Care During a Drought

If you simply have to mow your lawn during a drought, you can help the grasses by raising the cut level so that their leaves are left longer, as a low cut will weaken them. A long cut will also leave the grasses better placed to recover after rain. Also, don't collect the clippings. Instead, leave them on the surface as a mulch to reduce water evaporation from the soil, but be careful that the clippings themselves are not too long. If they are, they will potentially smother (and cause damage to) the lawn rather than mulching to its benefit.

A summer drought often ends with a heavy downpour, which can lead to flooding, waterlogging, and compaction. In these conditions, stay off the wet grass to prevent compaction.

Warmer temperatures mean a longer growing season. The downside is that lawns have to be mowed for longer, from early spring to late autumn, and in the mildest areas lawns may have to be mown all year-round. Weeds will benefit too. Moss likes to grow in spring before the grass becomes active, so warmer winters may result in moss growing for longer, but the grass will start to grow earlier too, so this may be enough to counter the moss growth.

Wet Winter Care

Lawns suffer just as much, if not more, in winter. Heavy rain combined with heavy foot traffic can leave the lawn a muddy mess, especially for those with less well-drained soils, clay, or compacted ground. The wet conditions encourage mosses, lichens, and algae to thrive. While the soil is waterlogged, stay off it to avoid further damage. A quick fix is to overseed the damaged areas. Once it stops raining and the soil

Ornamental grasses are a great replacement for lawns.

has dried enough to be worked and has warmed up enough for seed to germinate, rake the surface and expose the soil a little, sprinkle over seed, and press in lightly. Job done!

There are remedial actions you can take if your soil is prone to waterlogging. In autumn, you can improve it by forking, aerating, or slitting. This can also be done after waterlogging and flooding, but wait for the water to drain away.

A very poorly drained lawn may not recover from flooding because the standing water penetrates into the soil, pushing out the oxygen and creating anaerobic conditions around the roots. It may be better to start fresh by improving the soil and either re-sod or seed.

Ornamental Grasses

You may get rid of the lawn, but there's no reason why you can't grow ornamental grasses, which are far more drought resilient than their lawn grass cousins. A wide range of grasses are suited to a hot dry border, such as Japanese blood grass (*Imperata cylindrica*), marram grass (*Ammophila* spp.), Mexican feather grass (*Stipa tenuissima*), New Zealand wind grass (*Stipa arundinac*ea), pheasant's tail grass (*Anemanthele lessoniana*), pampas grass (*Cortaderia selloana*), sedges (*Carex* spp.), and switch grass (*Panicum virgatum*).

What Can I Grow for a Warmer Summer?

Looking ahead, you can think subtropical and Mediterranean. One of the restrictions on the choice of plant is hardiness, so initially select those that are hardy and capable of surviving a cold winter. You could create a tropical feel with lush foliage and dense planting by selecting large-leaved foliage plants, such as dahlia and canna, but as the climate warms up, this type of planting can be supplemented with a greater range of subtropicals, such as agave, banana (*Musa* spp.), palm, ginger (*Hedychium* spp.), Indian mallow (*Abutilon* spp.), and even pineapple (*Ananas comosus*).

Protecting Tender Exotics

With extreme cold becoming less common, it may be easy to get complacent and not bother to protect the less hardy of your plants over winter. The best approach is to make sure any borderline exotics are

Tropical plants that are not fully cold hardy will need to be protected in winter conditions.

A dry, gravel garden with rows of lavender.

wrapped up for winter. Some plants can be surprisingly resilient; the hardy banana (*Musa basjoo*), for example, is pretty tough and it can grow back after losing its leaves to cold and frost due to having hardy roots. Cordylines also have resilient roots and will regrow from the base of the stem if the temperatures don't drop too low.

Tolerant of Water and Heat

If a plant is tolerant of a hot dry summer, it's likely that it won't like waterlogging and floods in winter, and this is an important consideration in the future. However, research into tolerance of Mediterranean species to flooding has found that some species are more tolerant than first thought and may be useful species for the climate change garden. They include lamb's-ear (*Stachys byzantina*), lavender (*Lavandula angustifolia*), rock rose (*Helianthemum* spp.), and sage (*Salvia* spp.). They were found to be able to tolerate up to seventeen days in floodwater, both in summer and winter.

Mediterranean Flora

Many parts of the world have a Mediterranean-like climate of hot, dry summers and mild, wet winters. As well as the Mediterranean Basin itself, this climate is found in California, southern and southwestern Australia, Mexico, central Chile, and South Africa, and each has its own characteristic assemblage of species adapted to the conditions. These plants tend to have silvery, needle-like or hairy leaves that reduce the surface area from which water can be lost and also reflect heat away from the leaf. The hairs help trap any early morning dew too. Often the leaves are aromatic due to an oily film that acts like a sunscreen. These plants usually have deep roots to reach down to the water table and survive drought.

The fleshy-leaved species have a different survival strategy—they store water in their leaves and stems—while others have storage organs, such as tubers, bulbs, or rhizomes, which enable the plant to survive underground during the dry season. Those typical of the dry slopes of the Mediterranean include achillea, artemisia, cerinthe, cistus, eryngium, lavandula, nepeta, phlomis, santolina, salvia, and thyme. From the arid habitats of North America come echinacea, helenium, penstemon, and rudbeckia, while the South African fynbos is home to agapanthus, crocosmia, dierama, kniphofia, and protea.

Creating Your Own Mediterranean Garden

As dry summers with high temperatures and more sunshine hours become increasingly common, along with the lessening risk of late frost, there is an opportunity for gardeners to grow plants more usual of the Mediterranean region.

What's Typical of a Mediterranean Garden?

Mediterranean gardens typically have gravel paths with a backdrop of formal shapes created from trees and clipped hedges, contrasted with the irregular planting of drought-tolerant plants. Shade is key, both for people and plants to survive the heat of the day, so imagine pergolas or trellises creating shady corners with seating surrounded by scrambling plants, such as vines, hops, jasmine, or a scented rambling rose. Water, too, is typical of this style of garden, not just for the visual and sound effects, but also to create humidity and provide wildlife with a source of water.

Mediterranean and subtropical plants need a sunny spot and well-drained soil, so ground preparation is important. Although we are great fans of the no-dig method (see page 74), you need to get the soil right at the start, so mix in organic matter to create the essential well-drained bed. But the soil doesn't need much by way of fertility, as this will cause the plants to put on too much lush growth that is more susceptible to winter damage. The aim is to boost moisture retention and drainage. If you'd like, you can mulch with gravel to conserve as much water as possible. The gravel will also reflect heat, helping to keep the roots cool in the hot summer and suppress weeds. It also helps to keep water away from the base of the stems, which reduces the risk of rotting in winter.

Terra-cotta pots create a typical Mediterranean look and allow you to grow specimens that can cope with high summer temperatures but need protection from the cold and wet in winter. Here are some other ideas for plantings.

Lavender loves the heat in summer.

Specimen lemon trees in large pots.

Succulents and herbs are surrounded by gravel.

Type of Planting	Specific Plants
Statement plants	Agapanthus, agave, canna, Chusan palm (*Trachycarpus fortunei*), cordyline, Ethiopian banana (*Ensete ventricosum*), fan palm (*Chamaerops humilis*), fig (*Fatsia japonica*), New Zealand flax (*Phormium* spp.), phoenix palm (*Phoenix canariensis*), and yucca. The more risky species can be planted in large containers unless you live in a frost-free zone, but some of the hardier ones can stay outdoors all year.
Aromatic, sun-loving evergreen herbs	There are many to choose from, including bay (*Laurus nobilis*), lavender (*Lavandula* spp.), rosemary (*Salvia rosmarinus*), sage (*Salvia officinalis*), and thyme (*Thymus* spp.). Plant them near seating and path edges so visitors brush up against them and release the scent.
Shrubs	Options include bottlebrush (*Callistemon* spp.), broom (*Spartium junceum*), ceanothus, choisya, cistus, euphorbia, euryops, hebe, hibiscus, laurustinus, *Lomelosia minoana*, olearia, osmanthus, perovskia, *Retama sphaerocarpa*, and santolina.
Sun-loving herbaceous perennials	Acanthus, achillea, anchusa, artemisia, centaurea, echinops, echium, eryngium, hylotelephium, kniphofia, lamb's-ear, lavatera, phlomis, salvia, and valeriana.
Grasses to soften the borders	Mexican feather grass (S*tipa tenuissima*), miscanthus, New Zealand wind grass (*Agrostis avenacea*), and switch grass (*Panicum virgatum*).
Climbing plants	Creeping fig (*Ficus pumila*), crimson glory vine (*Vitis coignetiae*), grape vine (*Vitis* spp.), honeysuckle (*Lonicera* spp.), Japanese wisteria (*Wisteria floribunda*), trumpet vine (*Campsis* spp.) (some are hardy), and winter jasmine (*Jasminum nudiflorum*).
Plants for pots	Agave, aloe, bay tree (*Laurus nobilis*), bougainvillea, lantana, lemon, olive (*Olea europaea*), osteospermum, and pelargonium.
Trees	Italian cypress (*Cupressus sempervirens*), Italian privet (*Ligustrum* spp.), Portuguese laurel (*Prunus laurocerasus*), and stone pine (*Pinus pinea*).
Bromeliads	Fascicularia and puya.

RIGHT: **Naturalistic plantings consisting of plants native to your part of the world may prove more resilient in the face of climate change. This garden in Pennsylvania, USA is filled with plants native to the area.**

About the Authors

Sally Morgan is a botanist with a lifelong interest in gardening. She's written articles and books on food, farming, and the environment and owns an organic farm in Somerset, where she teaches courses on small farming (empirefarm.co.uk). Sally is the editor of *Organic Farming* magazine and gives talks on various garden- and farm-related subjects across the United Kingdom.

Kim Stoddart is a homesteader and writer. She has been covering climate change and resilient, savvy growing for publications such as *The Guardian*, since 2013. She edits *The Organic Way* magazine and writes for many outlets including *Grow Your Own* and *Country Smallholding* magazines. In addition, Kim runs online and in person courses from her climate change training gardens in Wales and around the U.K. (greenrocketcourses.com).

Other Books by the Authors

The Healthy Vegetable Garden
by Sally Morgan
(Chelsea Green, 2021)

Living on One Acre or Less
by Sally Morgan
(Green Books, 2016)

The Primrose Water Feature Book
by Kim Stoddart
(IPN, 2018)

Index

D

E

F

G

H

I

R

S

T

U

V

W

X

Y